Electronics
Second level

Hutchinson TECtexts

Hutchinson
TECtexts

Electronics

Second Level

G. Billups & M. T. Sampson

Hutchinson of London

Hutchinson & Co. (Publishers) Ltd
3 Fitzroy Square, London W 1 P 6JD

London Melbourne Sydney Auckland
Wellington Johannesburg and agencies
throughout the world

First published 1978

© G. Billups and M. T. Sampson 1978
Illustrations © Hutchinson & Co. (Publishers) Ltd 1978

Filmset in Times New Roman by Thomson Litho Ltd
East Kilbride, Scotland

Illustrations drawn by Oxford Illustrators Ltd

Printed in Great Britain by Thomson Litho Ltd
and bound by Wm Brendon & Son Ltd

ISBN 0 09 133331 8

Contents

Introduction

In each of the books in this series the authors have written text material to specified objectives. Test questions are provided to enable the reader to evaluate the objectives. The solutions or answers are given to all questions.

Topic area Elementary theory of semiconductors

Section 1   Semiconductor materials

After reading the following material, the reader shall:

1   Understand the simple concept of semiconductors.

1.1   Define the properties of semiconductors in relation to conductors and insulators.

1.2   State the two common types of semiconductor material as silicon and germanium.

1.3   Recognize the crystal structure of intrinsic (pure) silicon and germanium.

Conduction in any material is dependent upon its atomic structure. Electrons (called valence electrons) in the outer shell of the atom possess a high energy content and are less tightly bound to the atom than electrons orbiting in inner shells. The interlocking of valence electrons from two or more neighbouring atoms forms stable structures of the substance. Any electrons in the outer shell which are not used for valence (or 'structural linking') purposes are called 'free' electrons (sometimes 'conduction' electrons) and are free to move at random within the material.

Conductors (most of the metals) contain a large number of 'free' electrons. When an e.m.f. is applied across the ends of a conductor as shown in Figure 1(i), electrons drift in a definite direction, producing an electric current. The low resistance and low resistivity (or high conductivity) of conductors is a consequence of these facts. At the other end of the scale are the insulators (PVC, wood, glass, etc.), which have very few 'free' electrons and consequently current does not flow easily, if at all, through them. Insulators therefore possess a high resistance and high resistivity (or low conductivity). Somewhere between the good conductor and the perfect insulator is the semiconductor, of which silicon and germanium are the most common. Silicon and germanium both possess four valence electrons in their outer shell. These are shared with one valence electron from each of four neighbouring atoms (see Figure 2) to form a stable crystal structure which acts as an insulator at Absolute Zero. However, at room temperature (approximately 20°C) some of these valence

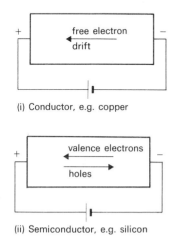

(i) Conductor, e.g. copper

(ii) Semiconductor, e.g. silicon

Figure 1 Conduction in (i) Conductor (ii) Semiconductor.

electrons break away from their atoms and a small electric current can flow as shown in Figure 1(ii). Consequently, although the resistance and resistivity of the semiconductor are high, they are not as high as the perfect insulator. The conductivity is also low but not as low as the perfect insulator.

After reading the following material the reader shall:

1.4   State that extrinsic semiconductor material can be formed by doping intrinsic silicon and germanium with impurities.

1.5   State that n-type semiconductor material is formed by the addition of a pentavalent material.

1.6   State that p-type semiconductor material is formed by the addition of a trivalent material.

1.7   Explain simply the structure of p-type and n-type semiconductors.

Pure silicon and germanium (called 'intrinsic' silicon and germanium) have four valence electrons in their outer shell. These four valence electrons are shared with one valence electron from each of four neighbouring atoms to form four covalent bonds as shown in Figure 2.

This crystal structure is very stable except at high temperature, or when an e.m.f. is applied large enough to result in the physical breakdown of the crystalline structure.

The poor conductivity of intrinsic germanium and silicon can be increased by deliberately introducing a controlled amount of impurity. This process is known as *doping*. The degree of doping is extremely small, of the order $1:10^5$ to $1:10^8$. Material with the proper amount of doping is called an *extrinsic* semiconductor. It has a conductivity somewhere between the high conductivity of a conductor and the low conductivity of an insulator. The impurities which are added are of two types:

(i)   A material which has five valence electrons in its outer shell, e.g. arsenic, antimony, phosphorus. Such material is said to be *pentavalent*. A typical atom is shown diagrammatically in Figure 3(i).

(ii) A material which has three valence electrons in its outer shell, e.g. boron, aluminium, indium, gallium. Such material is said to be *trivalent*. A typical atom is shown diagrammatically in Figure 3(ii).

The addition of (i) produces an n-type extrinsic material.
The addition of (ii) produces a p-type extrinsic material.

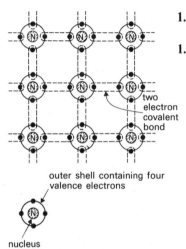

outer shell containing four valence electrons

nucleus
silicon or germanium atom (only the outer shell is shown for clarity)

Figure 2   Crystal structure of intrinsic silicon and germanium.

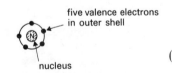

five valence electrons in outer shell

nucleus

(i) Pentavalent material, e.g. arsenic, antimony, phosphorus

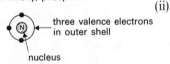

three valence electrons in outer shell

nucleus

(ii) Trivalent material, e.g. boron, aluminium, indium, gallium

Figure 3   (i) Pentavalent material
(ii) Trivalent material.

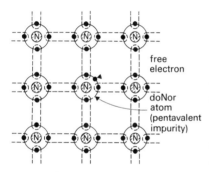

Figure 4 n-type material.

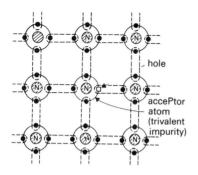

Figure 5 p-type material.

n-type material

The crystal structure of n-type material formed by the addition of a pentavalent material is shown in Figure 4. It can be seen that an extra electron exists for every impurity atom. The pentavalent impurity is responsible for donating this extra electron, and is hence termed a *donor* atom. This free electron is not tightly bound to the parent atom and can move randomly within the crystal structure. The motion of free electrons results in the movement of negative charges (since an electron possesses a negative charge). The electrons are called the *majority charge carriers* in n-type material. It is these majority charge carriers which contribute to current flow.

p-type material

The crystal structure of p-type material formed by the addition of a trivalent material is shown in Figure 5. It can be seen that the impurity atom produces an electron deficiency in one of its bonds. The trivalent impurity is responsible for this incomplete bond which can accept an electron from the outer shell of other atoms. The impurity atom is thus termed an *acceptor* atom. The electron deficiency, called a *hole*, can be considered as a positive charge, since it represents the absence of an electron which possesses a negative charge. In p-type material the holes are called the majority carriers, and these contribute also to current flow.

Note: Despite the labelling of these extrinsic materials as p and n-type the materials are still neutral, since they consist of both neutral intrinsic material and neutral impurity atoms. The p and n-type refer to the majority carriers within the doped material.

After reading the following material, the reader shall:

1.8 Explain electrical conduction as movement of electrons in n-type semiconductor material and 'apparent' movement of holes in p-type semiconductor material.

Conduction in n-type material

Consider a piece of extrinsic n-type semiconductor material with an e.m.f. applied across its ends, as shown in Figure 6. The positive terminal of the e.m.f. source gives rise to a positive charge at one end of the material, and the negative terminal a negative charge at the other end. The majority carriers (electrons) in the n-type material are attracted towards the positive terminal of the e.m.f. source (since like charges repel, unlike charges attract). Free electrons also move from the negative terminal of the e.m.f. source to the right hand side of the

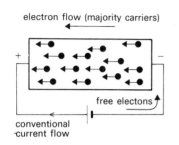

Figure 6 Conduction in n-type material.

semiconductor material. Within the semiconductor material, electrons (majority carriers) drift from the negative end to the positive end, and neutralize the positive charge of the impurity atoms left by electrons which have been moved to the conductor. Flow in the conductor is by free electrons which have been attracted to the positive terminal of the e.m.f. source. The complete circulating flow of electrons thus constitutes an *electric current* in the direction shown in Figure 6.

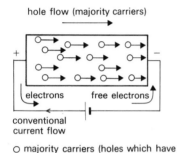

Figure 7   Conduction in p-type material.

Conduction in p-type material

An e.m.f. applied to extrinsic p-type semiconductor material gives rise to a positive charge at one end of the material, and a negative charge at the other end. The negatively charged end of the material attracts the majority carriers (holes) which provide a path for the free electrons, which move from the e.m.f. source negative terminal along the conductor. As current flow in the conductor is due to a drift of free electrons, holes do not move along the wire conductors. It is electrons from atoms at the left hand side of the extrinsic material which move to the e.m.f. source positive terminal along the wire conductor. Hence in p-type material the flow of electric current can be regarded as a drift of electrons in the direction shown in Figure 7.

After reading the following material, the reader shall:

1.9   State how a change in temperature affects the intrinsic conduction in a semiconductor.

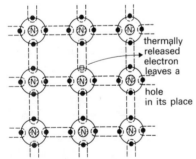

Figure 8   Intrinsic conduction in a semiconductor.

At room temperature thermal agitation of the atoms in intrinsic materials causes a few valence electrons to break away from the covalent bonds leaving a hole, as shown in Figure 8. Since one free electron always gives rise to one hole, the two are often referred to as a *hole/electron pair*. The presence of these hole/electron pairs increases the conductivity of the intrinsic material. Further increase in temperature above room temperature gives increased conductivity, until at approximately 80°C for germanium and 200°C for silicon the crystal structure breaks down.

If the temperature could be reduced to Absolute Zero, then there would be no thermal agitation or vibration of the atoms and all the valence electrons would be fixed in the atomic structure: the semiconductor has an insulator type of structure at this temperature. If the temperature increased, the structure would revert back to that of the semiconductor.

Thus conduction in an intrinsic semiconductor is partly due to a drift of valence electrons and partly due to a movement of holes. An increase in temperature causes more electron/hole pairs to be created, and hence conductivity increases as the temperature increases.

Self-assessment questions

1   Match the properties *A*, *B*, *C*, *D*, etc. with the two materials by placing the appropriate letter(s) in the space adjacent to the materials.

insulator	*A*	low resistance
	B	low resistivity
conductor	*C*	low conductivity
	D	high resistance
	E	high resistivity
	F	high conductivity

2   Which of the following materials are the most common intrinsic semiconductors? Underline the correct answer(s).

silica
aluminium
silicon
boron
germanium
phosphorus
indium

In questions 3 to 7, select the correct option.

3   An n-type material is formed by the addition of trivalent material to intrinsic semiconductor material.

TRUE/FALSE

4   A p-type material is formed by the addition of trivalent material to intrinsic semiconductor material.

TRUE/FALSE

5   In p-type material the majority carriers are ELECTRONS/HOLES.

6   In n-type material the majority carriers are ELECTRONS/HOLES.

7   Doping intrinsic silicon with arsenic produces p-type extrinsic semiconductor.

TRUE/FALSE

8   If the temperature of an intrinsic semiconductor material is increased the number of hole/electron pairs within the material

(*a*)   increases
(*b*)   decreases
(*c*)   remains the same

Underline the correct answer.

9   List two trivalent materials used in the formation of p-type material.

(*a*)

(*b*)

10   List three pentavalent materials used in the formation of n-type material.

(*a*)

(*b*)

(*c*)

Section 2   The semiconductor diode and its applications

After reading the following material, the reader shall:

2   Know the behaviour of a p–n junction with forward or reverse bias.
2.1   State how a barrier potential is formed across a p–n junction.
2.2   State that the barrier potential may be represented as a virtual battery.

Figure 9(i) represents a p–n junction, formed when extrinsic p- and n-type materials (with equal concentrations) are fused together. Owing to their random movement, holes in the p-type material in the vicinity of the junction diffuse across into the n-type material. Similarly electrons from the n-type material diffuse across the junction into the p-type. Combination of these hole/electron pairs takes place and results in a region where there are no free electrons or holes. This region is known as the *depletion layer*; its width is of the order 10^{-7} metres. The region near the junction in the p-type material takes up an excess negative charge, and the region near the junction in the n-type material takes up an excess positive charge. This charge distribution is shown in Figure 9(ii). The congregation of the positive

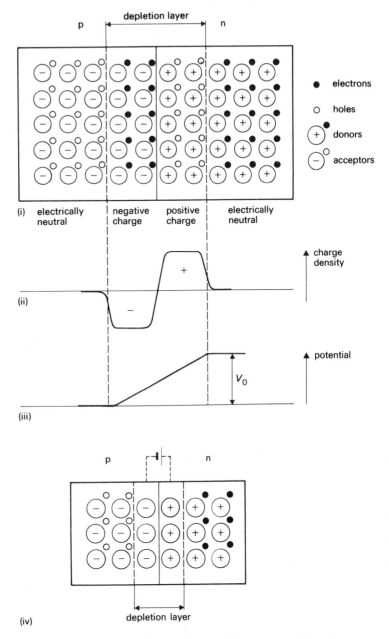

Figure 9 The formation of a barrier potential across a p–n junction.

and negative charges gives rise to a *barrier potential* called V_0, as shown in Figure 9(iii). The barrier may be represented as an imaginary battery as shown in Figure 9(iv). Typical barrier potentials are 0·3 V for germanium and 0·6 V for silicon.

After reading the following material, the reader shall:

2.3 Draw a p–n junction connected in the reverse bias mode, indicating current flow in the diode and the external circuit.

2.4 State why current does not flow in a p–n junction connected in the reverse bias mode.

2.5 Draw a p–n junction connected in the forward bias mode, indicating current flow in the diode and the external circuit.

2.6 State why current flows through a p–n junction connected in the forward bias mode.

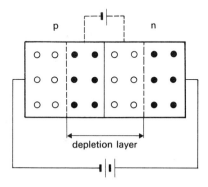

Figure 10   p–n junction connected in the reverse bias mode.

Consider Figure 10, showing the positive terminal of the e.m.f. source connected to the n-type material, and the negative terminal connected to the p-type material. This creates a negative charge on one side of the p-type material, and a positive charge at the other side. Consequently holes in the p-type material drift towards the negative charge, and electrons in the n-type material drift towards the positive

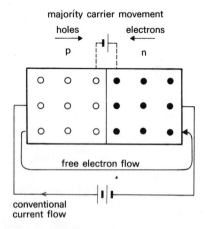

Figure 11   p–n junction connected in the forward bias mode.

charge. This creates a widening of the depletion layer, which has no majority current carriers (holes or electrons) present. Consequently very little current flows through the p–n junction and the external circuit. The p–n junction is now said to be *reverse biased*. The small current which does flow is due to the hole/electron pairs created by thermal agitation in the depletion layer at room temperature.

Consider Figure 11, showing the positive terminal of the e.m.f. source connected to the p-type material, and the negative terminal connected to the n-type material. Holes from the p-type material and electrons from the n-type material drift towards the junction, reducing the width of the depletion layer and hence also reducing the barrier potential. At the junction, holes and electrons combine to form hole/electron pairs: for every hole/electron pair combination at the junction, one electron and one hole flow across the junction, resulting in a current in the external circuit. The p–n junction is now said to be *forward biased*.

After reading the following material, the reader shall:

2.7 Sketch a test circuit diagram for determining the forward and reverse bias characteristics of a p–n junction diode.

2.8 Measure the current flow through a p–n junction connected in the forward bias mode.

2.9 Measure the current flow through a p–n junction connected in the reverse bias mode.

2.10 Sketch the static characteristic for a diode.

2.11 Compare the junction potentials of germanium and silicon diodes when connected in the forward bias mode.

2.12 Compare typical static characteristics for germanium and silicon diodes to illustrate the difference in forward voltage drop and reverse current.

A ammeter (to read mA in the forward bias mode, to read μA in the reverse bias mode)

V valve voltmeter

Figure 12 Circuit for determining the forward and reverse bias characteristics of silicon and germanium diodes.

The above objectives would be carried out in the laboratory using the typical test circuit shown in Figure 12. The results obtained would be plotted in graph form to show the forward and reverse static characteristics for both silicon and germanium diodes.

A typical static characteristic of a p–n junction diode is shown in Figure 13. In the forward bias mode, a small p.d. needs to be applied across the junction before any current flows: the barrier potential across the junction must be overcome before the diode will conduct.

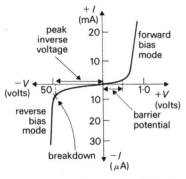

Figure 13 Static characteristic of a diode.

This barrier potential is typically 0.3 V for germanium and 0.6 V for silicon (see Figure 9). Once this voltage has been reached, very little change in the p.d. is required to produce a large current through the diode. The resistance of the diode in the forward bias mode is therefore low.

In the reverse bias mode very little current flows as the voltage is increased, until a point is reached when breakdown occurs, and considerable current flows. Breakdown for most diodes simply means the diode is permanently damaged. The value of the voltage at which breakdown occurs is referred to as the *peak inverse voltage* (p.i.v.) of the diode. The resistance of the diode in the reverse bias mode is high.

Table 1 illustrates typical results obtained in the laboratory using the test circuit of Figure 12. The silicon diode used was a 1N5041 (Radio Spares) and the germanium diode was an OA86 (Mullard). A graphical analysis of the results from this table can be seen in Figure 14, for the silicon diode. Using the results of Table 1, plot the characteristics for the germanium diode on the same axes as the silicon diode characteristic.

Table 1 Forward and reverse characteristics of silicon and germanium diodes.

germanium				silicon			
forward		reverse		forward		reverse	
V(Volts)	I(mA)	V(Volts)	I(μA)	V(Volts)	I(mA)	V(Volts)	I(μA)
0	0	0	1·0	0·41	0	0	0
0·40	1·0	1	1·1	0·48	0·2	2	1·0
0·52	2·0	2	1·5	0·51	0·5	4	1·7
0·64	3·0	4	2·0	0·525	0·7	6	2·25
0·72	4·0	5	2·2	0·53	1·0	8	2·90
0·80	5·0	10	2·5	0·58	2·0	10	3·40
0·88	6·0	12	2·8	0·60	4·0	14	4·50
0·96	7·0	14	3·0	0·62	6·0	16	5·10
1·03	8·0	16	3·5	0·64	8·0	18	5·80
1·10	9·0	18	4·0	0·65	10·0	20	6·60
1·16	10·0	20	5·0				

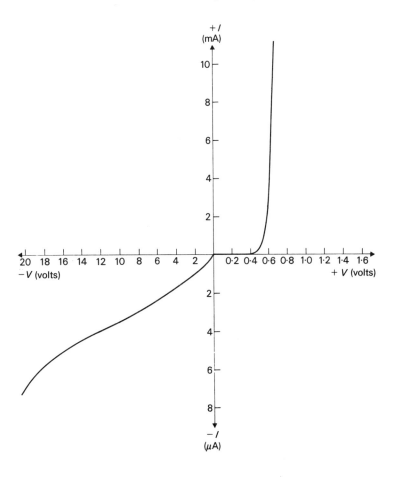

Figure 14 Forward and reverse characteristics of silicon and germanium diodes.

After reading the following material, the reader shall:

2.13 Explain the importance of considering peak inverse voltage of the diode.

2.14 Identify the breakdown effect.

The peak inverse voltage of a diode is the *maximum* safe value of voltage which can be applied to the diode in the reverse bias mode before breakdown occurs. This is of particular importance in circuits where the diode is used as a rectifier, and may have to withstand a reverse voltage which is considerably greater than the r.m.s. supply voltage.

The reverse current of the diode is important, because beyond breakdown the voltage across the diode remains constant over a wide range of reverse current. A diode which can be operated in the breakdown region without damage to the diode is called *a zener diode*. Also, if the diode is operated as a switch it is important to keep this reverse current to a minimum.

Figure 15   Reverse bias characteristics demonstrate the breakdown effect.

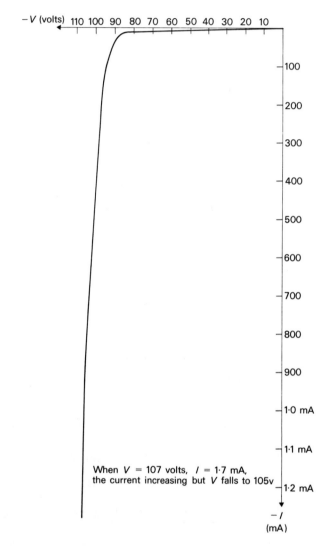

When V = 107 volts, I = 1·7 mA, the current increasing but V falls to 105v

Using the circuit of Figure 12 the reverse characteristics for a germanium gold bonded diode (Mullard AAY30) were obtained. The characteristics can be seen in Figure 15. Observe from the characteristics that at 107 volts the current continues to increase for no further increase in the reverse voltage. It is at this point that the diode is just beginning to break down.

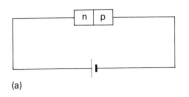

(a)

Figure 16 (a) p–n junction.

Self-assessment questions

1 In the diagram of Figure 16(*a*) the p–n junction is reverse biased.

TRUE/FALSE

2 To forward bias a p–n junction, the negative terminal of the e.m.f. source is connected to the n-type/p-type material. Underline the correct alternative.

3 When a p–n junction is forward biased, the majority carriers (holes in the p-type and electrons in the n-type) move towards the junction reducing the width of the depletion layer, hence causing the barrier potential also to be reduced.

TRUE/FALSE

4 When a p–n junction is reverse biased, the majority carriers (holes in the p-type and electrons in the n-type) move towards/away from the junction. Underline the correct alternative.

5 On the axes in Figure 16 (*b*) sketch the typical forward and reverse characteristics for a *silicon* p–n diode. Indicate on the diagram the forward and reverse bias modes.

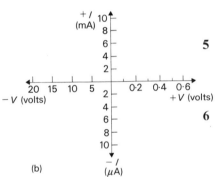

(b)

Figure 16 (b) Axes for forward and reverse characteristics of a silicon diode.

6 The forward bias junction potential for a silicon diode is approximately 0·6 V.

TRUE/FALSE

7 The forward bias junction potential for a germanium diode is approximately 1·5 V.

TRUE/FALSE

8 Any semiconductor diode can be operated within the breakdown region without irreversible damage.

TRUE/FALSE

9 The maximum reverse bias voltage which can be applied to a non-conducting silicon diode without irreversible damage is called:

(*a*) the zener voltage
(*b*) the peak inverse voltage
(*c*) the cut off voltage
(*d*) the minimum inverse voltage

Underline the correct answer.

10   Statement 1: A silicon diode can withstand a much higher reverse voltage than a germanium diode.
Statement 2: The reverse current of a germanium diode is considerably less than that of a silicon diode.

(a)   Only statement 1 is true.
(b)   Only statement 2 is true.
(c)   Both statements 1 and 2 are true.
(d)   Neither statement 1 nor 2 is true.

Underline the correct answer.

After reading the following material, the reader shall:

3   Know simple applications of semiconductor devices.
3.1   State simple applications of the available range of:

(a)   power diodes,
(b)   zener diodes,
(c)   signal diodes.

(a)   *Power diodes*

The largest application of the p–n junction is as a rectifying device. At the present time, ratings of power diodes for use as rectifying devices now exceed hundreds of amperes at higher than 1 kV peak inverse voltage, with a maximum reverse leakage current of 15 mA.

(b)   *Zener diodes*

The zener diode has its largest application as voltage stabilizers in power supply units. Zener diodes are available with voltage reference values from 1 V to 300 V with maximum power ratings of up to 100 watts. Other applications include meter protection circuits, instrument scale expanders, etc.

(c)   *Signal diodes*

These diodes, as their name implies, are for use with small amplitude signals. They are used in detector and mixer circuits and in logic gate circuits. Some can be used at high frequencies and are therefore used as frequency multipliers, which are used in radio and television systems.

After reading the following material, the reader shall:

3.2   Draw the circuit diagram required to produce half wave rectification using a p–n junction diode.

3.3   State the operation of the circuit required to produce half wave rectification using a p–n junction diode.

3.4   Draw the circuit diagram required to produce full wave rectification using p–n junction diodes connected in a bi-phase formation.

3.5   State the operation of the circuit required to produce full wave rectification using p–n junction diodes connected in a bi-phase formation.

3.6   Draw the circuit diagram required to produce full wave rectification using p–n junction diodes connected in bridge formation.

3.7   State the operation of the circuit required to produce full wave rectification using p–n junction diodes connected in bridge formation.

As its name implies, a half wave rectifier allows current to flow through the load on alternate half cycles of the a.c. input. The rectifier only conducts when *A* is positive with respect to *B* (see Figure 17), i.e.

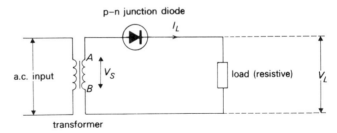

Figure 17   Half wave rectification circuit using a p–n junction diode.

during the positive half cycle of the a.c. input. The diode is forward biased during this half cycle. During the negative half cycle of the a.c. input, when *B* is positive with respect to *A*, the diode is reverse biased and does not conduct (except for the small leakage current). The output voltage across the load is a poor approximation to a steady d.c. voltage. It is a unidirectional pulsating voltage. (See Figure 20 for waveforms.)

As the name implies, full wave rectification allows current flow through the load during both half cycles of the a.c. input.

Figure 18 shows a full wave bi-phase circuit using a centre tapped transformer. The dots on the transformer windings indicate the ends of the windings which are at the same potential, i.e. when the dot end

Solutions to self-assessment questions (pages 21 and 22)

1   TRUE.

2   To forward bias a p–n junction the negative terminal of the e.m.f. source is connected to the n-type material.

3   TRUE.

4   When a p–n junction is reverse biased, the majority carriers (holes in the p-type and electrons in the n-type) move away from the junction.

5   See Figure 16 (*c*).

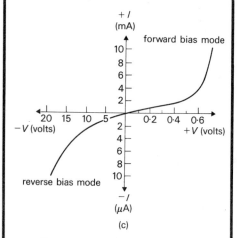

Figure 16 (c) Solution to Question 5.

6   TRUE.

7   FALSE. The forward bias junction potential for a germanium diode is approximately 0·3 V.

8   FALSE. Most diodes are damaged when the reverse current exceeds the breakdown point. Only the zener diode operates within the breakdown region without irreversible damage.

9   (*b*)   the peak inverse voltage.

10   (*a*)   Only statement 1 is true. (Statement 2 should read: The reverse current of a germanium diode is considerably greater than that of a silicon diode.)

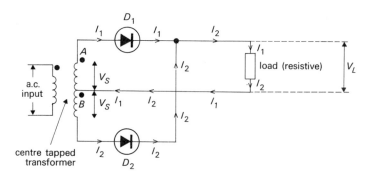

Figure 18   Full wave rectification circuit using p–n junction diodes connected in biphase formation.

of the transformer primary winding is at a positive potential, the corresponding ends of the two secondary windings are also positive. When point A is positive—i.e. during the positive half cycle of the a.c. input—diode D_1 is forward biased. Conduction takes place and current I_1 flows through diode D_1, the load, and to the centre tap of the transformer. Diode D_2 is reverse biased during the positive half cycle of the a.c. input, and does not conduct (except for the small leakage current). When point B is positive—i.e. during the negative half cycle of the a.c. input—diode D_2 is forward biased. Conduction takes place, and current I_2 flows through diode D_2, the load, and to the centre tap of the transformer. Diode D_1 is reverse biased during the negative half cycle of the a.c. input and does not conduct (except for the small leakage current).

The output voltage across the load is a unidirectional pulsating voltage employing both half cycles of the a.c. input voltage. (See Figure 21 for waveforms.)

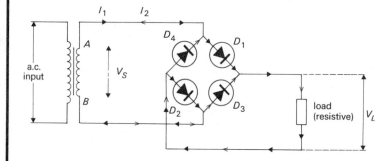

Figure 19   Full wave rectification circuit using p–n junction diodes connected in bridge formation.

Figure 19 shows a full wave rectification circuit using p–n junction diodes connected in bridge formation. When point A is positive with respect to B, diodes D_1 and D_2 conduct, diodes D_3 and D_4 being reverse biased. Current I_1 therefore flows from point A through diode D_1, the load and diode D_2 to point B. When point B is positive with respect to A, diodes D_3 and D_4 conduct, diodes D_1 and D_2 being reverse biased. Current I_2 therefore flows from point B through D_3, the load and D_4 to point A.

The output voltage across the load is once again a unidirectional pulsating voltage employing both half cycles of the a.c. input voltage. (See Figure 22 for waveforms.)

After reading the following material the reader shall:

3.8   Sketch waveforms of applied a.c. voltage and load current for diode circuits which provide half wave and full wave rectification into a resistive load.

Half wave rectification

Figure 20 shows the waveforms obtained from the circuit of Figure 17 for half wave rectification. In practice the voltage across the load V_L, is always slightly less than the a.c. supply voltage V_S, due to the voltage dropped across the diode and the source resistance. (See Figure 20(i) and (ii).)

The voltage across the load V_L consists of alternate half sine wave pulses and zero half cycle pulses, as shown in Figure 20(ii). The current through the load I_L is in phase with the voltage across the load as shown in Figure 20(iii).

Figure 20(iv) shows the voltage across the diode during each half cycle. During the positive half cycle of the a.c. input, when the diode is forward biased, the voltage across it is quite small. During the negative half cycle of the a.c. input, when the diode is reverse biased, the voltage across the diode is almost equal to the magnitude of the supply voltage. (Some voltage is dropped across the source resistance.)

Full wave rectification

(*a*)   **Bi-phase circuit**
Figure 21 shows the waveforms obtained from the circuit of Figure 18 for full wave rectification. In practice the voltage across the load, V_L, is always slightly less than the a.c. supply voltage V_S, for the same

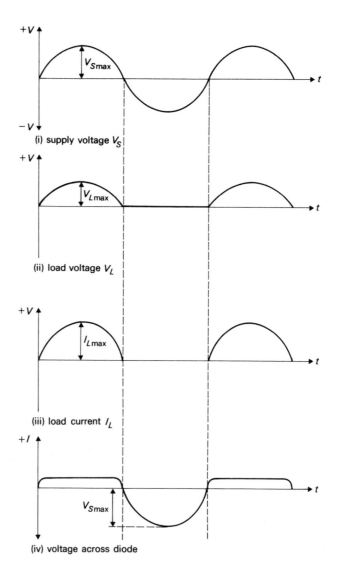

Figure 20   Waveforms from half wave rectification circuit.

reasons as for the half wave conditions. (See Figure 21(i) and (ii).) The voltage across the load consists this time of positive half sine wave pulses (the negative half cycles of the supply have been inverted) as shown in Figure 21(ii).

The current through the load I_L, consists of the currents I_1 and I_2; that is the current from diodes D_1 and D_2 respectively. I_L is again in phase with the voltage across the load. (See Figure 21(iii).)

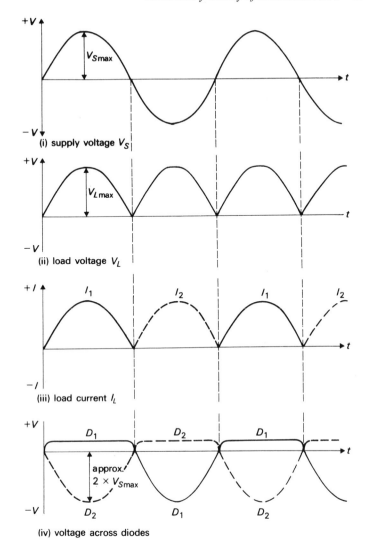

Figure 21 Waveforms from full wave rectification biphase circuit.

The voltage across diodes D_1 and D_2 is shown in Figure 21(iv). The peak inverse voltage across either diode is now twice the peak supply voltage; consequently for use in the bi-phase circuit diodes must be chosen which have a p.i.v. greater than twice the peak supply voltage.

(*b*) **Bridge circuit**
In practice the voltage across the load, V_L, is always slightly less than the a.c. supply voltage V_S, again for the same reasons as for the half wave conditions. (See Figure 22(i) and (ii).)

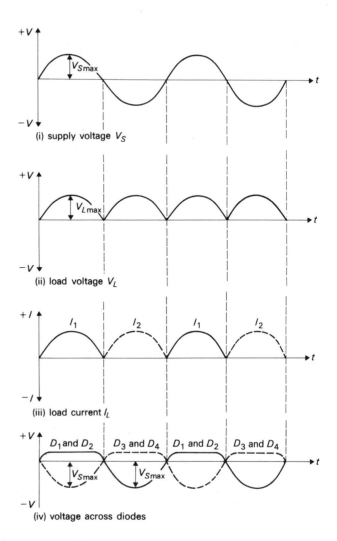

Figure 22   Waveforms from full wave rectification bridge circuit.

The voltage across the load consists of positive half sine wave pulses (the negative half cycles of the supply have been inverted) as shown in Figure 22(ii).

Again the current through the load, I_L, is in phase with the voltage across the load. (See Figure 22(iii).)

The diodes in the bridge circuit have only to withstand a p.i.v. of magnitude equal to the peak supply voltage $V_{S\,max}$, as shown in Figure 22(iv).

Note: The bridge circuit possesses two distinct advantages over the bi-phase circuit:

(i)   a centre-tap transformer is not required
(ii)   the diodes have only to withstand a p.i.v. of half the p.i.v. of diodes used in the corresponding bi-phase circuit.

After reading the following material, the reader shall:

3.9   Observe and measure the effects of connecting a smoothing capacitor across the load resistor in half and full wave rectifier circuits upon the diode current waveform, the load current waveform, the load p.d. waveform and the inverse voltage applied to the diode.

3.10   Draw the circuit diagram of a half wave rectifier incorporating a smoothing capacitor.

3.11   Draw the circuit diagram of a full wave rectifier (bi-phase circuit) incorporating a smoothing capacitor.

3.12   Draw the circuit diagram of a full wave rectifier (bridge formation) incorporating a smoothing capacitor.

The circuits of Figures 17, 18 and 19 produce a poor approximation to d.c. To give a closer approximation it is necessary to use additional components. The additional components required are called smoothing circuits. An electrolytic capacitor (sometimes called a reservoir capacitor) is the simplest of all smoothing circuits.

(i)   *Half wave rectification*

Using an oscilloscope, observations can be made of the (i) diode current waveform I_D, (ii) the load current waveform I_L, (iii) the load p.d. V_L, and (iv) the inverse voltage V_D applied to the diode for the half wave rectification circuit incorporating a smoothing capacitor, the circuit of Figure 23. Typical results observed by means of an oscilloscope can be seen in Figure 24.

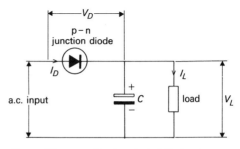

C smoothing capacitor (an electrolytic type).
Note the connections

Figure 23   Half wave rectifier with smoothing capacitor.

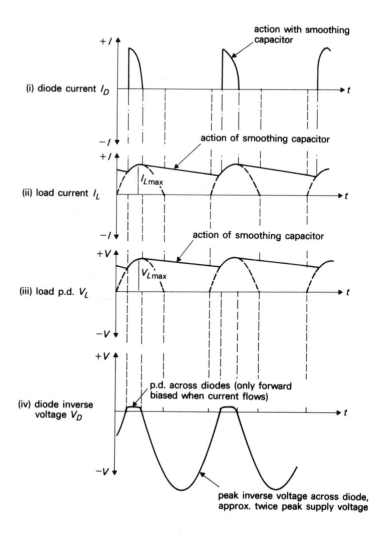

Figure 24   Waveforms obtained from the circuit of Figure 23.

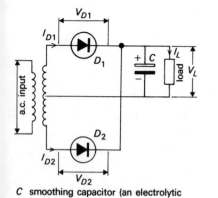

C smoothing capacitor (an electrolytic type)

Figure 25   Full wave rectifier (biphase) with smoothing capacitor.

(ii)   *Full wave rectification*

(a)   **Bi-phase circuit**
Using an oscilloscope observations can be made of the (i) diode current waveform I_{D1} and I_{D2}, (ii) the load current waveform $I_L = I_1 + I_2$, (iii) the load p.d. V_L, and (iv) the peak inverse voltage applied to the diodes V_{D1} and V_{D2} for the circuit of Figure 25. Typical results obtained in the laboratory can be seen in Figure 26.

(b)   **Bridge circuit**
Using an oscilloscope observations can be made of (i) the diode current waveform I_D, (ii) the load current waveform $I_L = I_1 + I_2$,

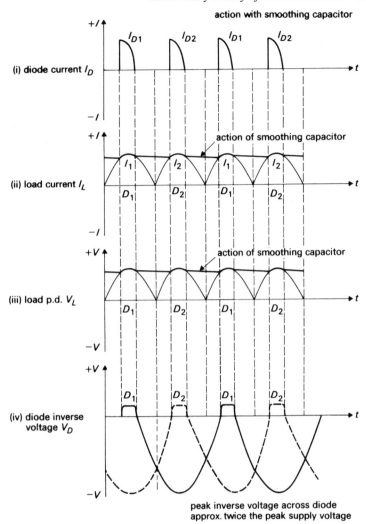

Figure 26 Waveforms obtained from the circuit of Figure 25.

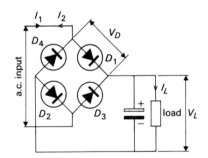

Figure 27 Full wave rectifier (bridge) with smoothing capacitor.

(iii) the load p.d. V_L, and (iv) the inverse voltage applied to the diode V_D for the circuit of Figure 27. Typical results obtained in the laboratory can be seen in Figure 28.

After reading the following material, the reader shall:

3.13 Sketch a circuit diagram of a stabilized voltage source including a zener diode and series resistor.

3.14 State the operation of the circuit which uses zener diode voltage stabilization.

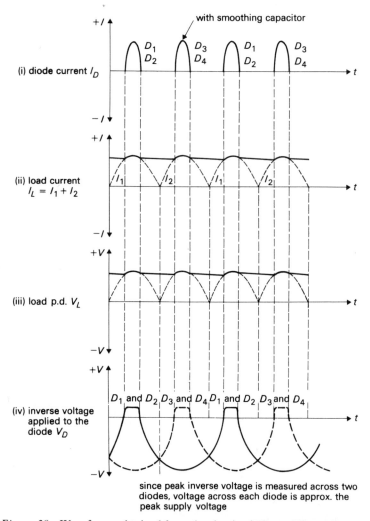

Figure 28   Waveforms obtained from the circuit of Figure 27.

3.15   Calculate the values of series resistor R in a simple zener stabilizing circuit for the conditions of:

(a)   varying supply voltage, fixed load,
(b)   fixed supply voltage, varying load.

The unstabilized output from the circuit of Figures 17, 18 and 19 is fed to the load resistor R_L, via a ballast resistor R_B and zener diode as shown in Figure 29. The output voltage across the load will remain constant for (i) change in load current and (ii) change in supply voltage.

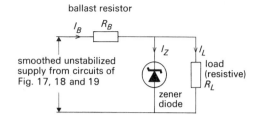

Figure 29 Circuit employing voltage stabilization using a zener diode.

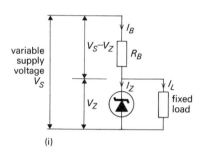

(i)

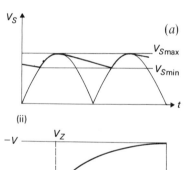

(ii)

(iii)

-I
(mA)

Figure 30 Zener diode voltage stabilization circuit: varying supply voltage, fixed load.

If the current in the load, I_L, increases the current in the zener diode I_Z falls, allowing the current in the ballast resistor, I_B, to remain constant. The output voltage therefore remains constant.

If the supply voltage increases, the current in the zener diode increases, causing the voltage across the ballast resistor to increase. The output voltage therefore remains constant.

Care must be taken to ensure that the current through the zener diode does not increase beyond the breakdown point or permanent damage to the zener diode will result.

(a) *Varying supply voltage, fixed load*

As shown in Figure 30(ii) the supply voltage V_s varies between V_{smax} and V_{smin}

Applying Kirchoff's first law to Figure 30(i)

$$I_B = I_L + I_Z$$

I_Z is a minimum when V_s is a minimum

∴ Value of ballast resistor required,

$$R_B = \frac{V_{smin} - V_Z}{I_B} \, \Omega$$

$$= \frac{V_{smin} - V_Z}{I_L + I_{Zmin}} \, \Omega$$

Consider Figure 30(iii). It is good practice to operate the zener diode just above the 'knee' of the characteristic and not near to the zero current condition. The zener voltage V_Z is then almost constant.

(b) *Fixed supply voltage, varying load*

As shown in Figure 31 the load current varies between I_{Lmax} and I_{Lmin}

I_Z is a minimum when the load current is a maximum.

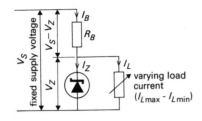

Figure 31   Zener diode voltage stabilization circuit: fixed supply voltage, varying load.

Applying Kirchoff's first law to Figure 31

$$I_B = I_{L\max} + I_{Z\min}$$

∴ Value of ballast resistor required,

$$R_B = \frac{V_s - V_Z}{I_B} \Omega$$

$$= \frac{V_s - V_Z}{I_{L\max} + I_{Z\min}} \Omega$$

Once again it is good practice to operate the zener diode just above the 'knee' of the characteristic as shown in Figure 30(iii).

Example

A 6·2 V zener diode is to be used as a stabilizer to supply a resistive load, which draws a load current of 30 mA. The minimum zener current to avoid the knee of the characteristic is 5 mA. The supply voltage varies between 12 and 15 V. Calculate the value of series resistor required in the stabilizing circuit.

Using the equation

$$R_B = \frac{V_{s\min} - V_Z}{I_L + I_{Z\min}} \Omega$$

$$= \frac{12 - 6\cdot2}{(30 \times 10^{-3}) + (5 \times 10^{-3})} \Omega$$

$$= \frac{5\cdot8}{35 \times 10^{-3}} \Omega$$

$$= \underline{\underline{165\Omega}}$$

Preferred value $= 180\Omega$ with 10% tolerance.

Example

A 9·1 V zener diode is to be used as a stabilizer to supply a resistive load whose current varies between 15 mA and 20 mA. If the supply voltage is fixed at 15 V and the minimum zener current is 1 mA, what value of series resistor is required in the stabilizing circuit?

Using the equation

$$R_B = \frac{V_s - V_Z}{I_{L\max} + I_{Z\min}} \Omega$$

$$= \frac{15 - 9\cdot1}{(20 \times 10^{-3}) + (1 \times 10^{-3})} \Omega$$

$$= \frac{5\cdot9}{21 \times 10^{-3}}$$

$$= \underline{\underline{281\Omega}}$$

Preferred value $= 270\Omega$ with 10% tolerance.

Self-assessment questions

11   Match the applications of diodes labelled *a* to *g* to the types of diodes labelled 1, 2 and 3 by placing the letters *a* to *g* next to the type of diode used.

1	zener diode	(*a*)	rectifying circuit
		(*b*)	voltage reference circuit
2	signal diode	(*c*)	mixer circuit
		(*d*)	frequency multiplier circuits
3	power diode	(*e*)	instrument scale expander
		(*f*)	logic gate circuits
		(*g*)	meter protection circuits

12   Figure 32 shows a group of isolated components. Make the interconnections necessary for the circuit to provide half wave rectification.

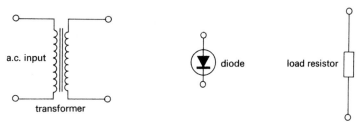

Figure 32   Components for half wave rectification circuit.

13   Sketch the expected waveforms of the a.c. input voltage, the voltage across the load and the current through the diode for the circuit of Question 2.

14   Sketch a circuit using the following components, which will provide full wave rectification.

> 4 power diodes
> 1 step down transformer
> 1 load resistor

15   In a bi-phase, full wave rectification circuit the peak inverse voltage across either of the diodes is approximately the same as the peak supply voltage.

TRUE/FALSE

16   In a full wave bridge rectification circuit the peak inverse voltage across any of the four diodes is approximately the same as the peak supply voltage.

TRUE/FALSE

17   State the simplest of all smoothing circuits which is additional to a rectifying circuit and is used to give a close approximation to d.c.

18   Sketch a circuit diagram of a bi-phase rectification circuit which uses a smoothing capacitor. Include the correct polarity of the smoothing capacitor on the circuit diagram.

19   Figure 25 shows a bi-phase rectifying circuit using a smoothing capacitor. For the two cases

(i)   without smoothing capacitor
      and
(ii) with smoothing capacitor,
      indicate the expected wave form (by letter) at the points in the circuit listed in the table.

points in the circuit	without capacitor	with capacitor
diode current waveform load current waveform load voltage waveform voltage across diode		

Expected wave forms

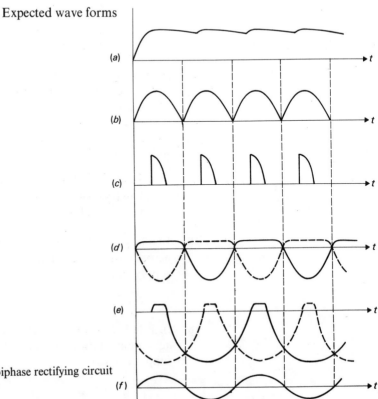

Figure 33 Expected waveforms for a biphase rectifying circuit with and without smoothing capacitor.

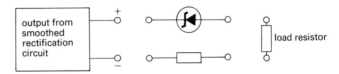

Figure 34 Components for constant load voltage.

20 Figure 34 shows a group of isolated components. Make the necessary interconnections in order that the load voltage remains constant, irrespective of any load current and supply voltage variations.

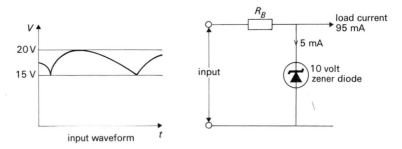

Figure 35 Zener diode circuit: varying supply voltage.

21 Calculate the value of series resistor (R_B) in the circuit of a simple zener diode voltage stabilizer, shown in Figure 35.

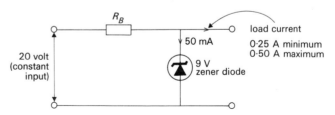

Figure 36 Zener diode circuit: varying load current.

22 Calculate the value of series resistor R_B required in the circuit of a simple zener diode voltage stabilizer, shown in Figure 36.

38 *Electronics: Second Level*

Solutions to self-assessment questions (pages 35–7)

11   1 zener diode   *b, e, g*
     2 signal diode   *c, d, f*
     3 power diode   *a*

12

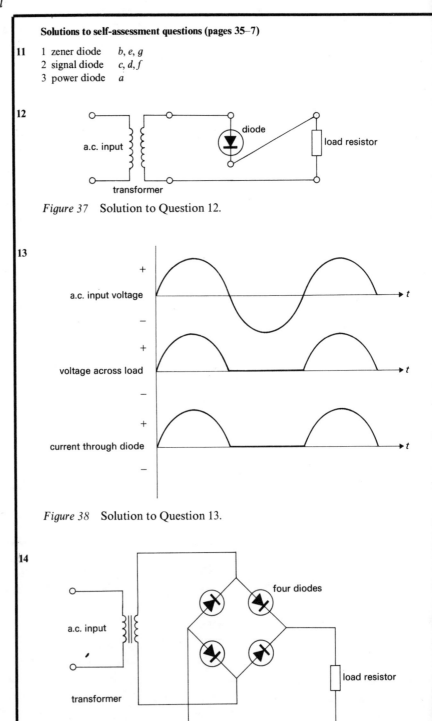

Figure 37   Solution to Question 12.

13

Figure 38   Solution to Question 13.

14

Figure 39   Solution to Question 14.

15 FALSE. Diodes for use in a bi-phase full wave rectification circuit must be chosen so that their p.i.v. is greater than twice the peak supply voltage.

16 TRUE.

17 Smoothing capacitor—usually an electrolytic capacitor.

18

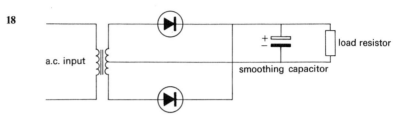

Figure 40 Solution to Question 18.

19

points in the circuit	without capacitor	with capacitor
diode current waveform	*b*	*c*
load current waveform	*b*	*a*
load voltage waveform	*b*	*a*
voltage across diode	*d*	*e*

20

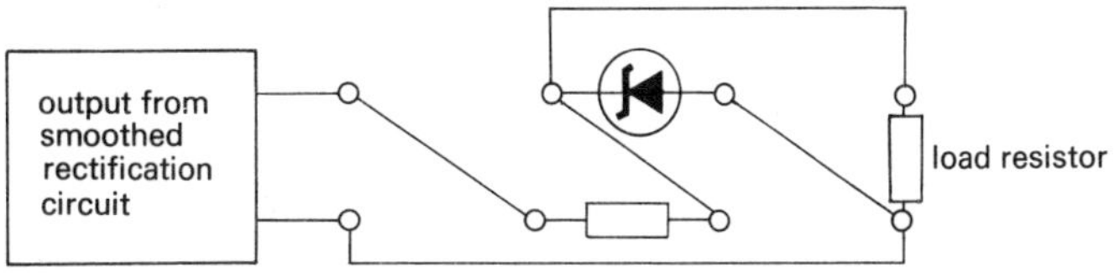

Figure 41 Solution to Question 20.

21

$$R_B = \frac{V_{smin} - V_Z}{I_L + I_{Zmin}} \Omega$$

$$= \frac{15 - 10}{95 \times 10^{-3} + 5 \times 10^{-3}} \Omega$$

$$= \frac{5}{100 \times 10^{-3}} \Omega$$

$$= \underline{\underline{50\Omega}}$$

22

$$R_B = \frac{V_s - V_Z}{I_{Lmax} + I_{Zmin}} \Omega$$

$$= \frac{20 - 9}{0\cdot50 + 0\cdot05} \Omega$$

$$= \frac{11}{0\cdot55} \Omega$$

$$= \underline{\underline{20\Omega}}$$

Section 3   The bipolar transistor

After reading the following material, the reader shall:

4 Know the arrangement of transistor electrodes.

4.1 Sketch the arrangement of a bipolar transistor produced from a sandwich of semiconductor materials.

4.2 Label the sections of the sketches for transistor types p–n–p, and n–p–n.

4.3 Identify the elctrodes of the bipolar transistor as emitter, collector and base.

4.4 Draw the circuit symbols for n–p–n and p–n–p bipolar transistors.

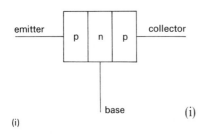

(i)

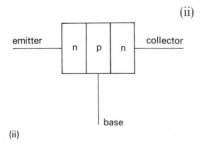

(ii)

Figure 42 The bipolar transistor, (i) p–n–p, (ii) n–p–n.

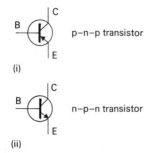

Figure 43 The transistor circuit symbol, (i) p–n–p, (ii) n–p–n.

The bipolar transistor as its name suggests employs both p-type and n-type semiconductor regions. The bipolar transistor is also commonly called the *junction* transistor. It is a three terminal device and two types exist:

(i) A layer of n-type material sandwiched between two layers of p-type material, and known as a p–n–p transistor (see Figure 42(i)).

(ii) A layer of p-type material sandwiched between two layers of n-type material, and known as a n–p–n transistor (see Figure 42(ii)).

Each of the three terminals of a transistor are designated the emitter, the base and the collector as shown in Figure 42(i) and 42(ii).

Transistors made with identical emitter and collector regions have poor performance. Performance may be improved if these regions are made of different physical size and conductivity. The emitter region is usually of low resistivity material which is heavily doped. The collector region has a slightly lower conductivity than the emitter region. The base region is usually of high resistivity material which is only very lightly doped. The junction formed between the emitter and base is the 'emitter' junction whilst the junction formed between the collector and base is the 'collector' junction.

Figure 43(i) shows the circuit symbol for a p–n–p transistor and Figure 43(ii) that for a n–p–n bipolar transistor. The difference between the two symbols is in the direction of the arrowhead on the emitter. It points towards the base in the p–n–p transistor, and away from the base in a n–p–n transistor. The direction of the arrowhead represents the direction of conventional current flow through the transistor.

Self-assessment questions

1

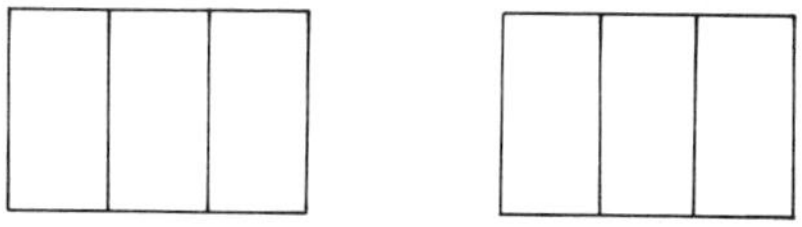

Figure 44 The two types of bipolar transistor.

Label the two sketches in Figure 44 to illustrate the two alternative types of bipolar transistor.

2 Supply the missing words.

In a bipolar transistor the emitter region is usually of _____ resistivity material which is _____ doped.

3 Select the appropriate word:

In a bipolar transistor the collector region is usually of lower/higher conductivity than the emitter region.

4 Supply the missing words:

In a bipolar transistor the base region is usually of _____ resistivity material which is only _____ doped.

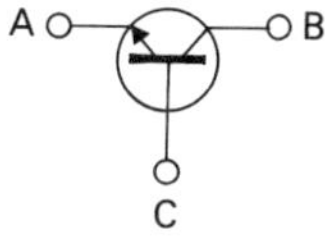

Figure 45 n–p–n transistor terminals.

5 Match the transistor terminals in Figure 45 of a n–p–n transistor to the names supplied.

emitter ( )
base ( )
collector ( )

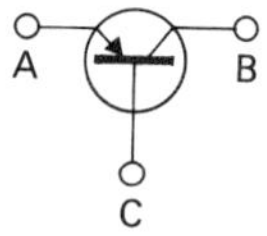

Figure 46 p–n–p transistor terminals.

6 Match the transistor terminals in Figure 46 of a p–n–p transistor to the names supplied.

emitter ( )
base ( )
collector ( )

Solutions to self-assessment questions (page 41)

1

Figure 47   Solution to Question 1.

2   In a bipolar transistor the emitter region is usually of LOW resistivity material which is HEAVILY doped.

3   In a bipolar transistor the collector region is usually of LOWER conductivity than the emitter region.

4   In a bipolar transistor the base region is usually of HIGH resistivity material which is only LIGHTLY doped.

5   emitter   (a)
      base   (c)
      collector (b)

6   emitter   (a)
      base   c)
      collector (b)

After reading the following material, the reader shall:

5   Know the modes of connection of a transistor.
5.1   Sketch the circuit diagrams for the common base, common emitter and common collector modes of connection.
5.2   Compare the relative values of input and output resistances for the three modes of connection.

When a transistor is to be connected in circuit, one of its terminals is the input terminal, one is the output terminal and the third is common to both input and output. Since the transistor is a three terminal device it may be connected in one of three modes:

(i)   the common base mode,
(ii)   the common emitter mode,
         and
(iii) the common collector mode.

Circuit diagrams of the three modes of connection are shown in Figure 48.

Table 2 shows typical values of input and output resistance for the three modes of connection of a transistor. Since input resistance is low in the common base mode and output resistance is high, this mode is

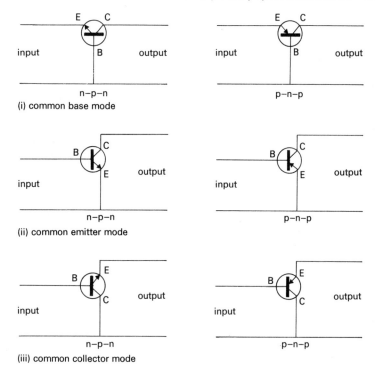

Figure 48 The three modes of connection of a transistor.

	common base	common emitter	common collector
input resistance	very low (50Ω–100Ω)	reasonably low (500Ω–2kΩ)	very high (500kΩ–1MΩ)
output resistance	very high 1MΩ	reasonably high 20kΩ	very low (50Ω–100Ω)

Table 2 Comparison of input and output resistance for the three modes of connection of a transistor.

usually employed to match a low resistance to a high resistance. The common emitter mode of connection is usually employed in amplifier circuits. The common collector mode of connection is usually employed in buffer stages to match a high resistance to a low resistance, since the input resistance is high in this mode and the output resistance low.

After reading the following material, the reader shall:

5.3 Draw the p–n–p and n–p–n transistor connected in the common base and common emitter modes indicating the flow of current through the device.

5.4 Define the short circuit current gains of a transistor connected in the common base (α) and the common emitter (β) modes.

5.5 State the relationship between α and β.

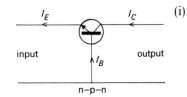

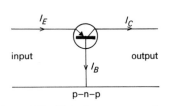

Figure 49   Transistor common base mode showing current flow.

(i)   *Common base*

Applying Kirchoff's first law to the circuits of Figure 49 for both the p–n–p and n–p–n transistor

$$I_E = I_C + I_B$$

The current gain of the transistor in the common base mode,

$$\alpha = \frac{\text{change in output current}}{\text{change in input current}}$$

i.e.

$$\alpha = \frac{\Delta I_C}{\Delta I_E}$$

where Δ is 'a small change in'

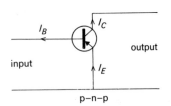

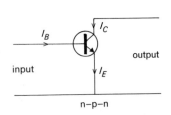

Figure 50   Transistor common emitter mode showing current flow.

(ii)   *Common emitter*

Applying Kirchoff's first law to the circuits of Figure 50 for both the p–n–p and n–p–n transistor

$$I_E = I_C + I_B$$

The current gain of the transistor in the common emitter mode,

$$\beta = \frac{\text{change in output current}}{\text{change in input current}}$$

i.e.

$$\beta = \frac{\Delta I_C}{\Delta I_B}$$

For any transistor I_C is always very much greater than I_B. If I_E is unity then I_C is approximately 0·98 and I_B is 0·02 (since $I_E = I_C + I_B$). Hence β in this case is 49. β usually lies between 50 and 99.

Note: It will be shown later that the values of α and β may also be calculated from the static characteristics of a transistor when connected in common base and common emitter modes.

Equation 1   As shown previously, $I_E = I_C + I_B$

and $\alpha = \dfrac{\Delta I_C}{\Delta I_E}$

Substituting from equation 1

$$\alpha = \frac{\Delta I_C}{\Delta I_C + \Delta I_B}$$

Inverting,

Equation 2

$$\frac{1}{\alpha} = \frac{\Delta I_C + \Delta I_B}{\Delta I_C}$$

$$= 1 + \frac{\Delta I_B}{\Delta I_C}$$

But

$$\frac{\Delta I_C}{\Delta I_B} = \beta$$

and

$$\frac{\Delta I_B}{\Delta I_C} = \frac{1}{\beta}$$

Substituting in equation 2,

$$\frac{1}{\alpha} = 1 + \frac{1}{\beta}$$

$$\frac{1}{\alpha} = \frac{\beta + 1}{\beta}$$

and

$$\alpha = \frac{\beta}{\beta + 1}$$

Similarly

$$\beta = \frac{\alpha}{1 - \alpha}$$

If $\alpha = 0.98$ then

$$\beta = \frac{0.98}{1 - 0.98} = \frac{0.98}{0.02} = 49$$

and if $\alpha = 0.99$ then

$$\beta = \frac{0.99}{1 - 0.99} = \frac{0.99}{0.01} = 99$$

As can be seen from these examples, a very small change in α corresponds to a large change in β.

Self-assessment questions

Complete the circuits below to obtain the three modes of transistor connection. Indicate clearly on each diagram the input and output terminals.

7

8

9

Figure 51 Common base mode.   *Figure 52* Common emitter mode.   *Figure 53* Common collector mode.

10 List the three modes of connection of a transistor in order of increasing input resistance.

(a)

(b)

(c)

11 List the three modes of connection of a transistor in order of increasing output resistance.

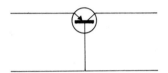

Figure 54 Common base mode current flow.

(a)

(b)

(c)

12 Given the p–n–p transistor of Figure 54 connected in the common base mode, indicate the direction of the currents in the circuit. Also label the currents.

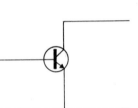

Figure 55 Common emitter mode current flow.

13 Given the n–p–n transistor of Figure 55 connected in the common emitter mode, indicate the direction of the currents in the circuit. Also label the currents.

14 The short circuit current gain, α, of a transistor connected in the common base mode is defined as

$$\alpha =$$

15 The short circuit current gain, β, of a transistor connected in the common emitter mode is defined as

$$\beta =$$

16 Which of the following relationships are true? Select the correct options.

(a) $\alpha = \dfrac{\beta}{\beta+1}$ TRUE/FALSE

(b) $\beta = \dfrac{\alpha}{\alpha+1}$ TRUE/FALSE

(c) $\alpha = \dfrac{\beta}{\beta-1}$ TRUE/FALSE

(d) $\beta = \dfrac{\alpha}{1-\alpha}$ TRUE/FALSE

After reading the following material the reader shall:

6 Know the static behaviour of a transistor.

6.1 Sketch a common base mode test circuit diagram for determining the static characteristics.

6.2 Describe the method of obtaining the common base mode static characteristics.

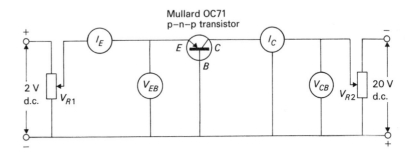

Figure 61 Circuit diagram for determining the static characteristics of a p–n–p transistor connected in the common base mode.

With the circuit connected as shown in Figure 61 the potentiometers V_{R1} and V_{R2} are set to zero. Potentiometer V_{R1} is adjusted to obtain a value of emitter current (input current) I_E of 2 mA. Potentiometer V_{R2} is adjusted to obtain 15 V across the collector base junction (V_{CB}). Maintaining I_E at 2 mA, V_{CB} is reduced by convenient amounts so that corresponding values of V_{EB} and I_C can be obtained. (Continue until $I_C = 0$ mA.) It will be necessary to reverse the potential of V_{CB} in order to reduce I_C to zero. The above procedure is then repeated with the emitter current I_E set for values of 4, 6 and 8 mA. Typical results obtained in the laboratory for a Mullard OC71 p–n–p transistor are shown in Table 3.

The potentiometers should again be set to zero. V_{R2} is adjusted to produce 0 V across the collector base junction, i.e. $V_{CB} = 0$ V. V_{R1} is then adjusted in suitable increments to obtain a set of readings for V_{EB}, I_E and I_C up to a maximum emitter current (I_E) of 10 mA. This procedure is then repeated for V_{CB} set at constant -5 V. Typical results obtained in the laboratory for a Mullard OC71 p–n–p transistor are shown in Table 4.

Note: If multirange instruments are used during the experiment it is important that their ranges should not be changed until a complete set of readings has been obtained.

7 Common base mode.

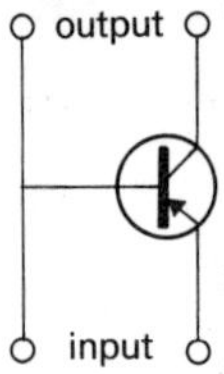

Figure 56 Solution to Question 7.

8 Common emitter mode.

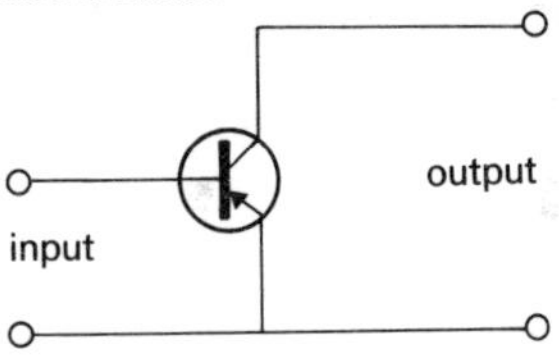

Figure 57 Solution to Question 8.

9 Common collector mode.

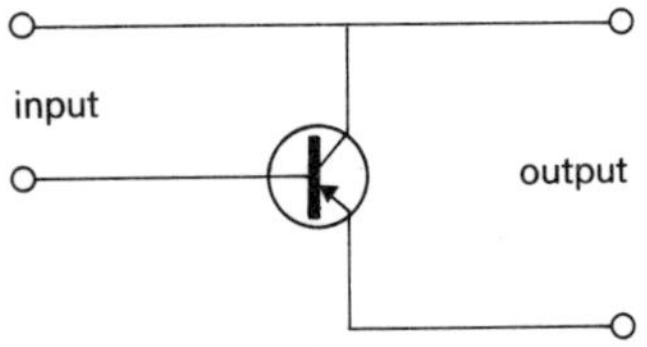

Figure 58 Solution to Question 9.

10 (a) Common base
(b) Common emitter
(c) Common collector

11 (a) Common collector
(b) Common emitter
(c) Common base

12

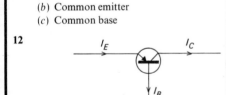

Figure 59 Solution to Question 12.

13

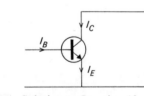

Figure 60 Solution to Question 13.

14 $\alpha = \dfrac{\text{change in output current}}{\text{change in input current}} = \dfrac{\Delta I_C}{\Delta I_E}$

15 $\beta = \dfrac{\text{change in output current}}{\text{change in input current}} = \dfrac{\Delta I_C}{\Delta I_B}$

16 (a) TRUE
(b) FALSE
(c) FALSE
(d) TRUE

Table 3 Transistor connected in the common base mode.

$IE = 2\,\mathrm{mA}$			$IE = 4\,\mathrm{mA}$		
I_C(mA)	V_{EB}(Volts)	V_{CB}(Volts)	I_C(mA)	V_{EB}(Volts)	V_{CB}(Volts)
2	0·12	15	3·92	0·14	15
2	0·12	10	3·95	0·14	10
2	0·13	5	3·91	0·16	5
1·97	0·14	1	3·85	0·185	1
1·95	0·15	0·25	3·83	0·18	0·25
0·12	0·32	0	0·28	0·4	0
0	0·32	−1	0	0·4	−1
$IE = 6\,\mathrm{mA}$			$IE = 8\,\mathrm{mA}$		
I_C(mA)	V_{EB}(Volts)	V_{CB}(Volts)	I_C(mA)	V_{EB}(Volts)	V_{CB}(Volts)
5·92	0·16	15	8·1	0·14	15
5·9	0·17	10	7·96	0·14	10
5·86	0·18	5	7·88	0·16	5
5·83	0·19	1	7·88	0·20	1
5·8	0·20	0·25	7·52	0·26	0·25
0·3	0·44	0	0·32	0·48	0
0	0·45	−1	0	0·49	−1

Table 4 Transistor connected in the common base mode.

$V_{CB} = 0\mathrm{V}$			$V_{CB} = -5\mathrm{V}$		
I_C(mA)	I_E(mA)	V_{EB}(Volts)	I_C(mA)	I_E(mA)	V_{EB}(Volts)
2	2·1	0·18	2	2·05	0·16
4	4·7	0·26	4	4·1	0·19
6	7·8	0·35	6	6·1	0·21
8	12	0·44	8	8·15	0·22
10	15	0·54	10	10·2	0·24

After reading the following material, the reader shall:

6.3 Plot typical curves of output, transfer and input characteristics for the transistor connected in the common base mode.

6.4 Discuss given typical families of curves of I_C/V_{CB} (output characteristics), I_C/I_E (transfer characteristics) V_{EB}/I_E (input characteristics).

Using the results of Tables 3 and 4, graphs of the output, transfer and input characteristics can be plotted as shown in Figures 62, 63 and 64 respectively.

Figure 62 shows the common base output characteristics which relate the variation in output current, I_C, and the output voltage V_{CB}. It can be seen that the collector current is almost constant for a wide range of collector/base voltage since the curves are almost horizontal. This emphasizes the high output resistance of the transistor in common base, since a large change in collector voltage produces only a very small change in collector current. It can also be seen that the collector current is maintained with almost zero collector voltage. This is due to the base current setting up a small potential which appears across

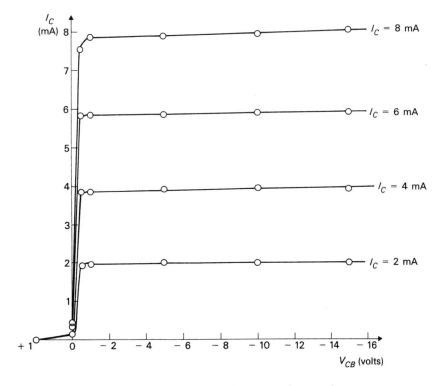

Figure 62 Output characteristics in common base mode.

the base collector junction, causing the collector junction to be reverse biased. Hence to reduce the collector current to zero it is necessary to apply a positive potential to the collector base junction. This is evident on the output characteristic; a potential of 1 V reduces the collector current to zero.

Figure 63 shows the common base transfer characteristics, which relate the input current I_E to the output current I_C. This characteristic is linear, i.e. its slope is constant: as I_E increases, I_C increases by a corresponding amount. The common base current gain of the transistor, α, can be calculated from this characteristic.

Figure 64 shows the common base input characteristics which relate the variation in input current, I_E, and the input voltage V_{EB}. This is the

Figure 63   Transfer characteristics in common base mode.

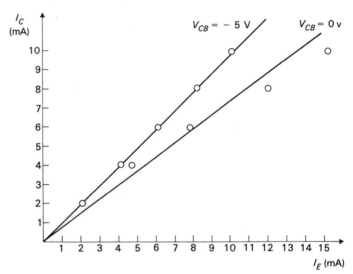

Figure 64   Input characteristics in common base mode.

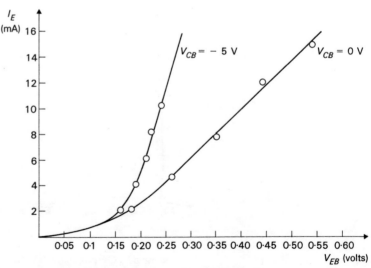

same characteristic as a forward biased p–n junction. At first, the emitter current increases only very slowly as the emitter voltage is increased. This is because the applied voltage needs to overcome the barrier potential of the base/emitter junction. Once the barrier potential has been neutralized, the emitter current increases rapidly, for only a very small change in the base/emitter voltage. This also emphasizes the low input resistance of the transistor when connected in the common base mode.

After reading the following material, the reader shall:

6.5 Sketch a common emitter mode test circuit diagram for determining the static characteristics.

6.6 Describe the method of obtaining the common emitter mode static characteristics.

$V_{CE} = -1 \cdot 0 \, \text{V}$

V_{BE} (Volts)	I_C (mA)	I_B (μA)
0·078	0·042	1·5
0·112	0·10	4
0·128	0·20	8
0·147	0·40	14·5
0·157	0·60	21
0·165	0·80	27
0·172	1·0	33

$V_{CE} = -6 \cdot 0 \, \text{V}$

I_C (mA)	I_B (μA)
0·10	3
0·20	7
0·40	12
0·60	17
0·80	22
1·0	29

Table 5 Transistor connected in the common emitter mode.

With the circuit connected as shown in Figure 65 the potentiometers V_{R1} and V_{R2} are set to zero. The power supplies are switched on, and the potentiometers adjusted until the collector/emitter voltage V_{CE} is -1 V. Potentiometer V_{R1} is then adjusted for a range of base current, I_B, and the corresponding values of V_{BE} and I_C recorded. Typical results obtained in the laboratory for a Mullard OC71 p–n–p transistor are shown in Table 5.

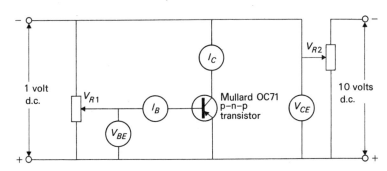

Figure 65 Circuit diagram for determining the static characteristics of a p–n–p transistor connected in the common emitter mode.

The potentiometers are now adjusted so that the base current I_B is 20 μA. For this fixed value of I_B, V_{R2} is now adjusted for a range of values of the collector emitter potential, V_{CE}, and the corresponding values of V_{CE} and I_C recorded. This procedure is repeated for fixed values of base current of 40 μA, 60 μA and 140 μA. Typical results obtained in the laboratory for a Mullard OC71 p–n–p transistor are shown in Table 6.

Table 6   Transistor connected in the common emitter mode.

V_{CE}(V)	$I_B = 20\mu A$	$I_B = 40\mu A$	$I_B = 60\mu A$	$I_B = 100\mu A$	$I_B = 140\mu A$
	I_C(mA)				
0·5	0·573	1·35	2·08	3·75	5·80
1·0	0·58	1·38	2·30	3·79	6·10
2·0	0·61	1·39	2·40	4·0	6·55
3·0	0·634	1·40	2·50	4·20	6·95
4·0	0·66	1·43	2·60	4·38	7·35
5·0	0·68	1·50	2·70	4·60	7·81
6·0	0·70	1·54	2·80	4·80	8·30
7·0	0·715	1·60	2·90	5·00	8·80
8·0	0·745	1·65	3·0	5·25	9·30
9·0	0·77	1·70	3·05	—	—
10·0	0·79	1·78	3·10	5·80	9·80

Note: If multirange instruments are used during the experiment it is important that their ranges should not be changed until a complete set of readings has been obtained.

After reading the following material, the reader shall:

6.7   Plot and describe typical families of curves of I_C/V_{CE} (output characteristics), I_C/I_B (transfer characteristic), V_{BE}/I_B (input characteristics).

Self-assessment question 17

Given the results of Tables 5 and 6, plot the graphs of the output, transfer and input characteristics.

After reading the following material, the reader shall:

6.8 Determine the values of α and β from given characteristics.

In an earlier section definitions of α and β were given as follows:

$$\alpha = \frac{\Delta I_C}{\Delta I_E} \quad \text{for constant } V_{CB}$$

$$\beta = \frac{\Delta I_C}{\Delta I_B} \quad \text{for constant } V_{CE}$$

α, the short circuit current gain of the transistor in the common base mode, can be calculated from the static transfer characteristic which is shown in Figure 63:

$$\alpha = \frac{\Delta I_C}{\Delta I_E}$$

$$= \text{slope of the common base static transfer characteristic}$$

By direct measurement from the common base static transfer characteristic of Figure 63,

$$\alpha = \frac{9 \cdot 6 \times 10^{-3}}{13 \times 10^{-3}} \quad \text{for } V_{CB} = 0 \text{ V}$$

$$\alpha = 0 \cdot 74$$

β, the short circuit current gain of the transistor in the common emitter mode, can be calculated from the static transfer characteristic which is shown in Figure 67:

$$\beta = \frac{\Delta I_C}{\Delta I_B}$$

$$= \text{slope of the common emitter static transfer characteristic}$$

By direct measurement from the common emitter static transfer characteristic of Figure 67,

$$\beta = \frac{0 \cdot 895 \times 10^{-3}}{30 \times 10^{-6}} \quad \text{for } V_{CE} = -1 \text{ V}$$

$$\beta = 29 \cdot 8$$

Clearly α is dependent upon V_{CB} and β is dependent upon V_{CE}, because the slope of the transfer characteristic in common base and common emitter is different for different values of V_{CB} and V_{CE}.

54   *Electronics: Second Level*

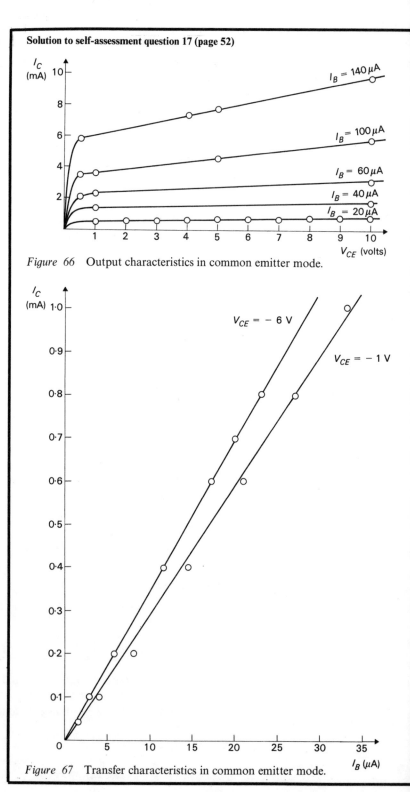

Figure 66 Output characteristics in common emitter mode.

Figure 67 Transfer characteristics in common emitter mode.

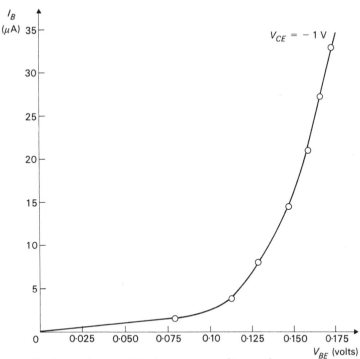

Figure 68   Input characteristics in common emitter mode.

The characteristics should be as in Figures 66, 67 and 68.

Figure 66 shows the common emitter output characteristics which relate the variation in output current I_C and the output voltage V_{CE}. This set of characteristics is very similar to the common base output characteristics (Figure 62) except that there is now a noticeable slope on each characteristic. This indicates that the output resistance is lower for the transistor connected in the common emitter mode than in the common base mode. Also notice that the collector current is zero for zero collector voltage. Unlike the common base mode the potential produced by the base current does not appear in the collector emitter circuit.

Figure 67 shows the common emitter transfer characteristics, which relate the input current I_B to the output current I_C. This characteristic is almost linear (except for very small currents). The common emitter current gain of the transistor, β, can be calculated from this characteristic.

Figure 68 shows the common emitter input characteristics which relate the variation in base current I_B as the base emitter voltage V_{BE} is increased in the forward direction. The input characteristic is a function of collector voltage, but the input resistance does not change significantly with the collector voltage. However, the input resistance does vary widely as the base voltage varies, since the characteristic is not linear.

After reading the following material, the reader shall:

6.9   Determine the value of input resistance from given input characteristics.

(i)   *Common base*

$$\text{Input resistance} = \frac{V_{EB}}{I_E} \quad \text{for constant } V_{CB}$$

= reciprocal of the slope of the common base static input characteristic

At low values of V_{EB} the input resistance is not constant, and the value of the input resistance is usually taken over the linear portion of the input characteristic.

By direct measurement from the common base static input characteristic of Figure 64,

$$\text{Input resistance} = \frac{0 \cdot 55 - 0 \cdot 135}{15 \cdot 6 \times 10^{-3}} \quad \text{for constant } V_{CB} = 0\,\text{V}$$

Input resistance = $26 \cdot 6\,\Omega$

(ii)   *Common emitter*

$$\text{Input resistance} = \frac{V_{BE}}{I_B} \quad \text{for constant } V_{CE}$$

= reciprocal of the slope of the common emitter static input characteristic

By direct measurement from the common emitter static input characteristic of Figure 68,

$$\text{Input resistance} = \frac{0 \cdot 175 - 0 \cdot 130}{34 \cdot 5 \times 10^{-6}} \quad \text{for constant } V_{CE} = 0\,\text{V}$$

Input resistance = $1 \cdot 3\,\text{k}\Omega$

It can be seen from the two calculations that the input resistance in the common base mode is considerably less than that in the common emitter mode, as previously shown in Table 2.

After reading the following material, the reader shall:

6.10   Determine the value of output resistance from given output characteristics.

(i) *Common base*

Output resistance $= \dfrac{V_{CB}}{I_C}$ for constant I_E

$\qquad\qquad\qquad\quad$ = reciprocal of the slope of the common base static output characteristic

By direct measurement from the common base static output characteristic of Figure 62,

Output resistance $= \dfrac{16-1}{(3 \cdot 98 - 3 \cdot 85) \times 10^{-3}}$ for constant $I_E = 4\,\text{mA}$

Output resistance $= 0 \cdot 123\,\text{M}\Omega$

(ii) *Common emitter*

Output resistance $= \dfrac{V_{CE}}{I_C}$ for constant I_B

$\qquad\qquad\qquad\quad$ = reciprocal of the slope of the common emitter static output characteristic

By direct measurement from the common emitter static output characteristic of Figure 66,

Output resistance $= \dfrac{10-1}{3 \cdot 1 - 2 \cdot 35}$ for constant $I_B = 60\,\text{mA}$

Output resistance $= 12\,\text{k}\Omega$

It can be seen from the two calculations that the output resistance in the common base mode is greater than that in the common emitter mode as shown previously in Table 2.

Self-assessment questions

18 When determining the common base static characteristics, the input and output currents and voltages must be measured. Complete each space with the symbols for:

$\qquad$ input current $\qquad$ ____
$\qquad$ input voltage $\qquad$ ____
$\qquad$ output current $\qquad$ ____
$\qquad$ output voltage $\qquad$ ____

19 Sketch the general shape of the following for a p–n–p transistor connected in the common base mode:

(i) Output characteristics
(ii) Transfer characteristics
(iii) Input characteristics

20 For each of the characteristics of Question 19 one quantity is constant. Complete each space with the symbol of the quantity.

(i)   Output characteristic   _____
(ii)   Transfer characteristic   _____
(iii) Input characteristic   _____

21 Statement 1: The output resistance of a transistor connected in the common base mode is high.

Statement 2: This is due to a large change in collector voltage producing a large change in collector current.

(a) Only statement 1 is true.
(b) Only statement 2 is true.
(c) Both statements 1 and 2 are true.
(d) Neither statement 1 or 2 is true.

Underline the correct answer.

22 The common base static transfer characteristic relating the input current I_E to the output current I_C is linear for constant collector base voltage.

TRUE/FALSE

23 Statement 1: The common base static input characteristic is similar to a reverse biased p–n junction.

Statement 2: Once the barrier potential of the base emitter junction has been neutralized the emitter current increases rapidly.

(a) Only statement 1 is true.
(b) Only statement 2 is true.
(c) Both statements 1 and 2 are true.
(d) Neither statement 1 or 2 is true.

Underline the correct answer.

24 When determining the common emitter static characteristics the input and output currents and voltages must be measured. Complete each space with the symbol of the quantity.

input current   _____
input voltage   _____
output current   _____
output voltage   _____

25 Sketch the general shape of the following for a p–n–p transistor connected in the common emitter mode:

(i)   Output characteristics

(ii)   Transfer characteristics
(iii) Input characteristics

26 For each of the characteristics of Question 8 one quantity is constant. Complete each space with the symbol of the quantity.

(i)   Output characteristic   _____
(ii)   Transfer characteristic   _____
(iii) Input characteristic   _____

27 Statement 1: The common emitter output characteristic is very similar to the common base output characteristic except with a more distinctive slope.

Statement 2: The output resistance is lower for the common emitter mode than for the common base mode.

(*a*) Only statement 1 is true.
(*b*) Only statement 2 is true.
(*c*) Both statements 1 and 2 are true.
(*d*) Neither statement 1 or 2 is true.

Underline the correct answer.

28 The common emitter static transfer characteristic relating the input current I_B to the output current I_C is almost linear except for very small currents.   TRUE/FALSE

29 Statement 1: The common emitter input resistance does change significantly with collector voltage.

Statement 2: The common emitter input resistance does change significantly with base voltage.

(*a*) Only statement 1 is true.
(*b*) Only statement 2 is true.
(*c*) Both statements 1 and 2 are true.
(*d*) Neither statement 1 nor 2 is true.

Underline the correct answer.

30 The short circuit current gain of a transistor, α, in the common base mode is defined as $\alpha = \Delta I_C/\Delta I_E$ for constant V_{CB}. How is α calculated from the static characteristics, and from which characteristic can it be found?

31 The short circuit current gain of a transistor, β, in the common emitter mode is defined as $\beta = \Delta I_C/\Delta I_B$ for constant V_{CE}. How is β calculated from the static characteristics and from which characteristic can it be found?

32   Statement 1: The input resistance in the common base mode is not constant, and is usually taken over the linear portion of the characteristic.

Statement 2: The input resistance in the common base mode can be calculated from the reciprocal of the slope of the common base static input characteristic.

(a)   Only statement 1 is true.
(b)   Only statement 2 is true.
(c)   Both statements 1 and 2 are true.
(d)   Neither statement 1 nor 2 is true.

Underline the correct answer.

33   Statement 1: The output resistance of a transistor connected in the common emitter mode is defined as:

$$\text{output resistance} = \frac{V_{CB}}{I_E} \quad \text{for constant } I_C.$$

Statement 2: The output resistance of a transistor connected in the common emitter mode can be found directly from the slope of the static output characteristic.

(a)   Only statement 1 is true.
(b)   Only statement 2 is true.
(c)   Both statements 1 and 2 are true.
(d)   Neither statement 1 nor 2 is true.

Underline the correct answer.

Post test – elementary theory of semiconductors

1 Classify the list of materials below into insulators, conductors and semiconductors by placing the letters I, C and S next to the material.

aluminium   silicon
germanium   brass
copper   PVC
mica   glass (molten)
glass (not molten)

2 List two electrical properties of a conductor.

3 List two electrical properties of an insulator.

4 At Absolute Zero a semiconductor behaves as a perfect insulator, whilst at room temperature it behaves as a perfect conductor.

TRUE/FALSE

5 Name the two types of extrinsic semiconductor material.

6 The type of impurity added in the formation of an n-type semiconductor material is a _____ material.

7 The type of impurity added in the formation of a p-type semiconductor material is a _____ material.

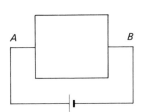

Figure 71 ń-type semiconductor with e.m.f. supplied.

8 The presence of hole/electron pairs in an intrinsic semiconductor will increase the conductivity of the intrinsic material.

TRUE/FALSE

9 The diagram of Figure 71 shows a piece of extrinsic n-type semiconductor material with an e.m.f. applied to it. Indicate the direction of electron drift between points *A* and *B*.

10 The diagram of Figure 72 shows a piece of extrinsic p-type semiconductor material with an e.m.f. applied to it. Indicate the direction hole drift between points *A* and *B*.

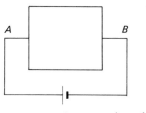

Figure 72 ṕ-type semi-conductor with e.m.f. supplied.

11 To forward bias a p–n junction, the negative terminal of the e.m.f. source is connected to the n-type/p-type material. Underline the correct alternative.

12 What are typical barrier potentials for

(i) Silicon p–n junction?
(ii) Germanium p–n junction?

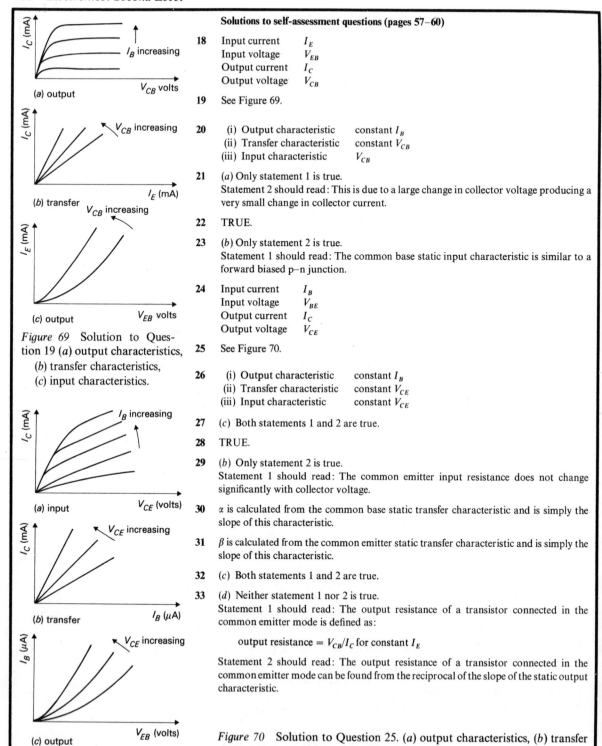

Figure 69 Solution to Question 19 (*a*) output characteristics, (*b*) transfer characteristics, (*c*) input characteristics.

Solutions to self-assessment questions (pages 57–60)

18   Input current I_E
      Input voltage V_{EB}
      Output current I_C
      Output voltage V_{CB}

19   See Figure 69.

20   (i) Output characteristic   constant I_B
      (ii) Transfer characteristic   constant V_{CB}
      (iii) Input characteristic V_{CB}

21   (*a*) Only statement 1 is true.
      Statement 2 should read: This is due to a large change in collector voltage producing a very small change in collector current.

22   TRUE.

23   (*b*) Only statement 2 is true.
      Statement 1 should read: The common base static input characteristic is similar to a forward biased p–n junction.

24   Input current I_B
      Input voltage V_{BE}
      Output current I_C
      Output voltage V_{CE}

25   See Figure 70.

26   (i) Output characteristic   constant I_B
      (ii) Transfer characteristic   constant V_{CE}
      (iii) Input characteristic   constant V_{CE}

27   (*c*) Both statements 1 and 2 are true.

28   TRUE.

29   (*b*) Only statement 2 is true.
      Statement 1 should read: The common emitter input resistance does not change significantly with collector voltage.

30   α is calculated from the common base static transfer characteristic and is simply the slope of this characteristic.

31   β is calculated from the common emitter static transfer characteristic and is simply the slope of this characteristic.

32   (*c*) Both statements 1 and 2 are true.

33   (*d*) Neither statement 1 nor 2 is true.
      Statement 1 should read: The output resistance of a transistor connected in the common emitter mode is defined as:

$$\text{output resistance} = V_{CB}/I_C \text{ for constant } I_E$$

      Statement 2 should read: The output resistance of a transistor connected in the common emitter mode can be found from the reciprocal of the slope of the static output characteristic.

Figure 70 Solution to Question 25. (*a*) output characteristics, (*b*) transfer characteristics, (*c*) input characteristics.

13 When a p–n junction is reverse biased, the majority carriers move away from/towards the junction. Underline the correct alternative.

14 Sketch the typical forward and reverse bias characteristics for a germanium p–n diode. Indicate typical scale magnitudes.

15 The maximum reverse bias voltage which can be applied to a non-conducting silicon diode without irreversible damage is called the

_____.

16 A small current flows in a reverse biased semiconductor diode.

TRUE/FALSE

17 A diode which operates in the reverse bias mode where the voltage across the diode remains constant over a wide range of reverse current is called a _____ diode.

18 In the circuit of Figure 73 label the diodes *A, B, C, D, E* from the given list. Each may be used once, more than once, or not at all.

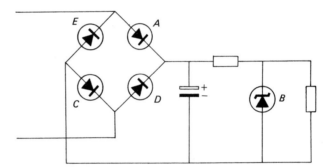

Figure 73 Circuit for Question 18.

(i) zener diode
(ii) signal diode
(iii) power diode

19 Sketch a circuit diagram which uses the following list of components to achieve full wave rectification.

2 power diodes
centre tapped transformer
load resistor

20 For the bi-phase full wave rectification circuit, sketch the expected waveforms of the a.c. input voltage, the load voltage, the load current and the voltage across each diode.

64 Electronics: Second Level

21 Statement 1: In a biphase full wave rectification circuit without smoothing, the peak inverse voltage across either of the two diodes is approximately the same as the peak supply voltage.

Statement 2: In a full wave bridge rectification circuit without smoothing, the peak inverse voltage across any of the four diodes is approximately the same as the peak supply voltage.

(a) Only statement 1 is true.
(b) Only statement 2 is true.
(c) Both statements 1 and 2 are true.
(d) Neither statement 1 nor 2 is true.

Underline the correct answer.

22 Sketch a half wave rectification circuit which incorporates a smoothing capacitor. Include the correct polarity of the smoothing capacitor on the circuit diagram.

23 Figure 74 shows a half wave rectification circuit employing a smoothing capacitor. For the two cases

(i) without smoothing capacitor
and
(ii) with smoothing capacitor, indicate the expected waveform by letter at the points in the circuit listed in the table below. Each waveform may be used once, more than once, or not at all.

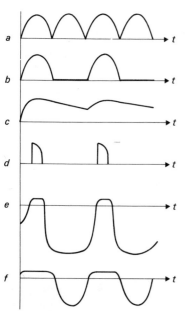

points in the circuit	without capacitor	with capacitor
diode current waveform load current waveform load voltage waveform voltage across diode		

Figure 74 Waveforms for Question 23.

24 Sketch a circuit diagram of a stabilized voltage source including a zener diode and a series resistor.

25 Label the sketches of Figure 75 to illustrate the two alternative types of bipolar transistor.

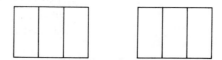

Figure 75 Two types of Bipolar Transistor (Question 25).

26   For each of the three regions of a bipolar transistor (emitter, base, and collector), write down the characteristics *a*, *b*, *c*, etc., of each region in the spaces provided. Each may be used once, more than once or not at all.

emitter	____	(*a*) lightly doped
base	____	(*b*) heavily doped
collector	____	(*c*) low resistivity
		(*d*) high resistivity
		(*e*) very lightly doped

27   Sketch the circuit symbol for a p–n–p transistor and mark on the three terminals their appropriate names.

28   Sketch the circuit symbol for a n–p–n transistor and mark on the three terminals their appropriate names.

29   Using the circuit symbol of a transistor show how the common base mode of connection might be achieved. Indicate clearly the input and output terminals.

30   Using the circuit symbol of a transistor show how the common emitter mode of connection might be achieved. Indicate clearly the input and output terminals.

31   List the three modes of connection of a transistor in order of decreasing output resistance:

(*a*)

(*b*)

(*c*)

32   List the three modes of connection of a transistor in order of decreasing input resistance:

(*a*)

(*b*)

(*c*)

33   Write down an equation relating the collector, base and emitter current for both a p–n–p and n–p–n transistor.

34   Define, using symbols, the short circuit current gain of a transistor connected in the common base mode.

35   Define, using symbols, the short circuit current gain of a transistor connected in the common emitter mode.

36   State a relationship between α and β.

37   When determining the common emitter static characteristics, the input current, input voltage, output current, and output voltage must be measured. Their symbols are:

(i)   Input current   ＿＿＿
(ii)   Input voltage   ＿＿＿
(iii) Output current   ＿＿＿
(iv)   Output voltage   ＿＿＿

38   Sketch a test circuit diagram for determining the static characteristics of a transistor connected in the common emitter mode.

39   The short circuit current gain of a transistor in the common base mode (α) can be obtained from the common base static transfer characteristic.

TRUE/FALSE

40   The short circuit current gain of a transistor in the common emitter mode (β) can be obtained from the common emitter static output characteristic.

TRUE/FALSE

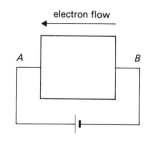

Figure 76 Solution to Question 9.

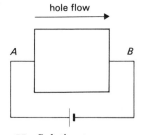

Figure 77 Solution to Question 10.

Solutions to post test (pages 61–66)

1

aluminium	C
germanium	S
copper	C
mica	I
glass (not molten)	I
silicon	S
brass	C
PVC	I
glass (molten)	C

2 (i) low resistance
(ii) low resistivity (high conductivity)

3 (i) high resistance
(ii) high resistivity (low conductivity)

4 FALSE

5 P- and n-type materials.

6 Pentavalent.

7 Trivalent.

8 TRUE.

9 See Figure 76.

10 See Figure 77.

11 N-type material.

12 (i) 0·6 V.
(ii) 0·3 V.

13 Away from.

14 See Figure 14, page 19.

15 Peak inverse voltage.

16 TRUE. A small leakage current flows due to hole/electron pairs created by thermal agitation.

17 Zener.

18 (i) Zener diode   *B*
(ii) Signal diode
(iii) Power diode   *A, C, D, E*

19 See Figure 18, page 24.

20 See Figure 21, page 27.

21 (*b*)

22 See Figure 23, page 29.

23 points in the circuit	without capacitor	with capacitor
diode current waveform	*b*	*d*
load current waveform	*b*	*c*
load voltage waveform	*b*	*c*
voltage across diode	*f*	*e*

24   See Figure 29, page 33.

25   See Figure 47, page 42.

26   emitter   *c, b*
     base   *d, e*
     collector   *a*

27   See Figure 43(i), page 40.

28   See Figure 43(ii), page 40.

29   See Figure 48(i), page 43.

30   See Figure 48(ii), page 43.

31   (*a*) Common base
     (*b*) Common emitter
     (*c*) Common collector

32   (*a*) Common collector
     (*b*) Common emitter
     (*c*) Common base

33   $I_E = I_C + I_B$

34   $\alpha = \dfrac{\Delta I_C}{\Delta I_E}$

35   $\beta = \dfrac{\Delta I_C}{\Delta I_B}$

36   $\alpha = \dfrac{\beta}{\beta + 1}$ or $\beta = \dfrac{\alpha}{1 - \alpha}$

37   (i) I_B
     (ii) V_{BC}
     (iii) I_C
     (iv) V_{CE}

38   See Figure 65, page 51.

39   TRUE.

40   FALSE. β is obtained from the common emitter static transfer characteristic.

Topic area Thermionic valves

After reading the following material, the reader shall:

7   Understand the concept of thermionic emission.

7.1   Describe the effect known as thermionic emission.

7.2   Define the space charge.

When a metal is heated (usually above room temperature) the random movement of the 'free' electrons is increased and some escape from the metal. The energy supplied as heat is given up to the electrons. Those escaping form a cloud of electrons above the metal surface known as a *negative space charge* (negative since an electron possesses a negative charge). This effect is known as *thermionic emission* and depends on two main factors:

(i)   the operating temperature

(ii)   the material used (usually a metal)

Common materials used for cathodes in thermionic devices include tungsten, thoriated tungsten and nickel which is oxide coated using a mixture of strontium and barium oxide. The oxide coating produces more electrons at a given temperature than does the tungsten or thoriated tungsten.

After reading the following material, the reader shall:

7.3   Label the diagram of a directly heated diode valve.

7.4   Identify the circuit symbol for a directly heated diode valve.

7.5   State the operation of a directly heated diode valve.

Figure 78(a) shows the detailed construction of a directly heated diode valve. Figure 78(b) shows its circuit symbol.

The diode valve is the simplest form of valve and as its name suggests contains only two electrodes—a nickel anode surrounding an oxide coated cathode. The cathode is directly heated by a current which flows through it and thermionic emission takes place producing a negative space charge which surrounds the cathode. Equilibrium is attained and the negative space charge limits further electron

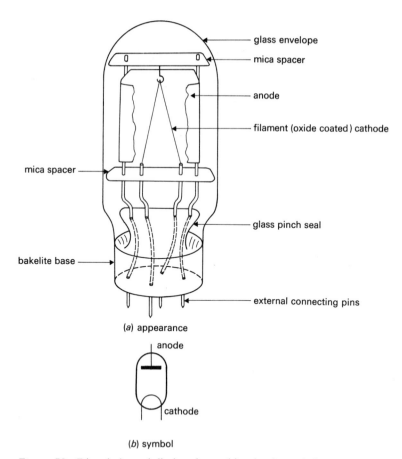

(a) appearance

(b) symbol

Figure 78   Directly heated diode valve and its circuit symbol.

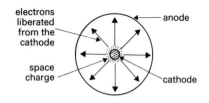

Figure 79   Electron flow in the diode valve.

emission from the cathode, i.e. the number of electrons emitted is counterbalanced by an equal number re-entering the cathode from the space charge.

If a positive potential is now applied to the anode with respect to the cathode, electrons (which have a negative charge) are attracted to the positive anode (since unlike charges attract). The electrons travel radially outwards from the cathode to anode as shown in Figure 79. This flow of electrons from anode to cathode is known as the *anode current*.

If the positive potential on the anode is further increased, more electrons are attracted to the anode and a point is reached when all the liberated electrons from the cathode reach the anode. This is known as the *saturation current*.

If a negative potential is applied to the anode with respect to the cathode, no current flows since the space charge electrons are repelled back to the cathode. Consequently *no* anode current will flow.

After reading the following material, the reader shall:

7.6   Label the diagram of an indirectly heated diode valve.
7.7   Identify the circuit symbol for an indirectly heated diode valve.
7.8   State the operation of an indirectly heated diode valve.

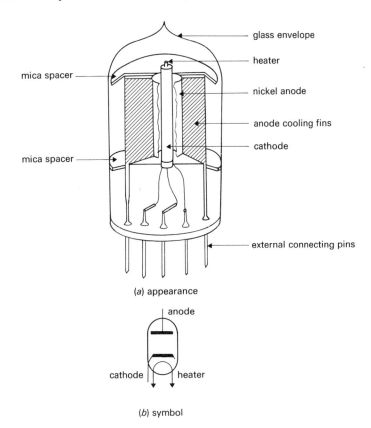

Figure 80   Indirectly heated diode valve and its circuit symbol.

Figure 80(a) shows the detailed construction of an indirectly heated diode valve. Figure 80(b) shows its circuit symbol.

The cathode of the indirectly heated diode valve is heated by a separate heating filament. As previously stated, the number of electrons which can be liberated from the cathode depends upon the temperature of the cathode metal. The cathode temperature of the directly heated diode valve is constant and hence the saturation current is fixed. This problem is overcome by the indirectly heated diode valve since its cathode temperature can be altered by changing the filament current. Consequently the indirectly heated diode valve can have a series of saturation currents each dependent on a filament current. Operation of this type of valve is identical to that of the directly heated type except that the cathode temperature can be changed.

After reading the following material, the reader shall:

8   Know the static behaviour of a thermionic diode.

8.1   Sketch a test circuit diagram for determining the anode characteristics of a thermionic diode.

8.2   State the method for obtaining the anode characteristic of a thermionic diode.

8.3   Plot and describe a typical anode characteristic.

8.4   Describe how current flow is produced in a thermionic diode by the correct application of a p.d. between the anode and cathode.

8.5   Explain the effect of 'saturation' in relation to anode characteristics.

8.6   Show the effect on anode characteristics of a change in heater current.

Figure 81 shows a typical circuit diagram for determining the anode characteristics of a Mullard GRD7 diode valve. With the circuit connected as shown in Figure 81 the filament current (I_F) was adjusted to 2 A and kept constant at this value. The anode voltage (V_A) was increased from zero to 200 V and the values of V_A and the corresponding value of anode current (I_A) were recorded at each increment. The same procedure was repeated for filament current of 2·1 A and 2·3 A. The results are shown in Table 7.

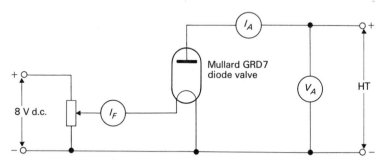

Figure 81   Circuit diagram for determining the anode characteristics of a thermionic diode.

Figure 82 shows the anode characteristics of a Mullard GRD7 diode valve, for various filament currents plotted from the results of Table 7. The characteristics show that as the anode voltage, V_A, is increased the anode current, I_A, increases almost linearly at first. This is due to the fact that not all the electrons leaving the cathode reach the anode due

$I_F = 2\,\text{A}$		$I_F = 2{\cdot}1\,\text{A}$		$I_F = 2{\cdot}2\,\text{A}$	
V_A (Volts)	I_A (mA)	V_A (Volts)	I_A (mA)	V_A (Volts)	I_A (mA)
20	4·85	20	6·2	20	8·5
40	5·1	40	9·2	40	16·5
60	5·2	60	9·325	60	17·25
80	5·275	80	9·4	80	17·5
100	5·3	100	9·425	100	18
120	5·325	120	9·5	120	18
140	5·5	140	10	140	18
160	5·525	160	10	160	18·1
180	5·675	180	10	180	18·25
200	5·7	200	10	200	18·75

Table 7 Typical results obtained for a Mullard GRD7 diode valve.

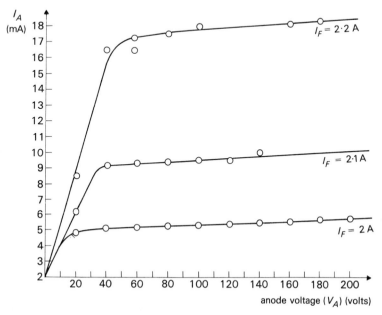

Figure 82 Anode characteristics of a Mullard GRD7 diode valve.

to the effect of the negative space charge around the cathode. Over this portion of the characteristic the relationship between anode current and anode voltage is given by the Langmuir–Child law, sometimes known as the three halves power law.

$I_A = kV_A^{3/2}$ where k is a constant involving the charge and mass of the electron and the electrode geometry

Further increase in anode potential causes all the electrons liberated from the cathode to reach the anode and the characteristic begins to flatten out. The diode is then said to be *saturated*, and any further increase in anode voltage cannot cause any further increase in anode current. Higher anode currents are possible, however, by increasing the filament current. This increases the cathode temperature, causing still more electrons to be liberated and giving a larger anode current.

After reading the following material, the reader shall:

9 Know and compare the rectifying action of diodes.

9.1 Compare the merits of thermionic and semiconductor diodes as rectifiers.

Comparison of the merits of thermionic and semiconductor diodes for use as rectifiers appears in the following table.

thermionic diodes	semiconductor diodes
1 Usually has a filament and therefore requires a separate supply. 2 Bulky. 3 No current flows on reverse bias. 4 Many leads in comparison to the semiconductor diode. 5 More expensive than the semiconductor diode. 6 Large d.c. supplies required.	1 No vacuum is required. 2 Only two leads (cathode and anode). 3 No filament. 4 Lower operating voltage. 5 Smaller in physical size than the equivalent diode valve. 6 Cheap. 7 Small current flows (leakage current) on reverse bias.

Self-assessment questions

Complete the following statements 1–6.

1 In thermionic valves the cloud of electrons around a heated cathode is called a _____.

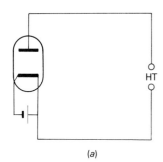

(a)

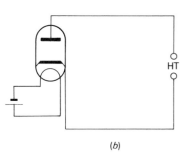

(b)

Figure 83 Diode valve method of heating.

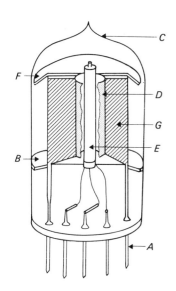

Figure 84 Indirectly heated valve.

2 In thermionic valves the process of liberating free electrons by means of heat is called _____.

3 Draw diagrams to represent the circuit symbols for

(i) a directly heated thermionic diode
(ii) an indirectly heated thermionic diode.

4 Thermionic emission depends on two factors. These are

(i) _____

(ii) _____

5 In the diagram of Figure 83(a), the method of heating the cathode is called _____.

6 In the diagram of Figure 83(b), the method of heating the cathode is called _____.

7 Figure 84 shows the constructional features of an indirectly heated thermionic diode valve. Match the component parts labelled A to G to the numbers 1 to 8.

A   1   heater
B   2   nickel anode
C   3   anode cooling fins
D   4   glass envelope
E   5   mica spacer
F   6   external connecting pin
G   7   cathode
                      8   control grid

8 Complete the following statement: When the current in a diode valve reaches the point where a further increase in voltage does *not* produce a further increase in current, the diode has reached _____.

9 Select the correct word.
The diode valve is reverse biased and will not conduct when the anode is positive/negative and the cathode is positive/negative.

10 When a diode valve is conducting the anode voltage is always positive with respect to the cathode.

TRUE/FALSE

11 Connect the apparatus of Figure 85 such that the static characteristic of a diode valve can be obtained.

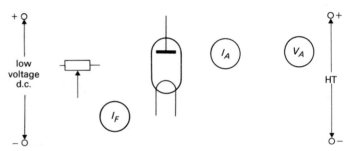

Figure 85   Components necessary to obtain the static characteristics of a diode valve.

12   Complete the following statement.
In a thermionic diode valve the anode current depends on

(*a*)   _____

(*b*)   _____

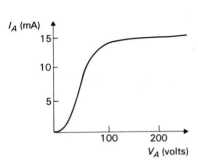

Figure 86   Static characteristic of a diode valve.

13   The graph of Figure 86 shows the static characteristic of a diode valve. State the saturation voltage.

The saturation voltage is _____ volts.

14   For a fixed anode voltage the anode current can be increased if the cathode temperature is:

(*a*)   decreased
(*b*)   increased
(*c*)   unchanged
(*d*)   reduced slightly

Tick the correct answer

15   List the features in this text which are true of:

(i)   thermionic diode
(ii)   semiconductor diode

(*a*)   anode
(*b*)   filament
(*c*)   no vacuum
(*d*)   many leads
(*e*)   high voltage
(*f*)   two leads
(*g*)   cathode
(*h*)   low voltage
(*j*)   bulky
(*k*)   no filament
(*l*)   compact
(*m*)   grid
(*n*)   cheap
(*p*)   leakage current

After reading the following material, the reader shall:

10   Know the static behaviour of a thermionic triode.
10.1 Label the diagram of a thermionic triode valve.
10.2 Sketch the circuit symbol of a thermionic triode valve indicating the anode, cathode, grid and heater.
10.3 State the operation of a thermionic triode valve.
10.4 Explain the effect of control grid voltage on anode current.

Figure 89(*a*) shows the detailed construction of a thermionic triode valve. Figure 89(*b*) shows its circuit symbol.

The triode valve has three electrodes—a cathode, an anode and a control grid. The control grid, in the form of a wire mesh, is placed between the anode and cathode. The potential on the control grid can be controlled. As it is closer to the cathode than the anode the variation in grid potential has more effect on the anode current than would the same anode potential. If there is zero potential on the control grid the triode valve operates as a diode valve. The anode is always made positive with respect to the cathode, and as in the diode valve, electrons from the cathode space charge are attracted to the

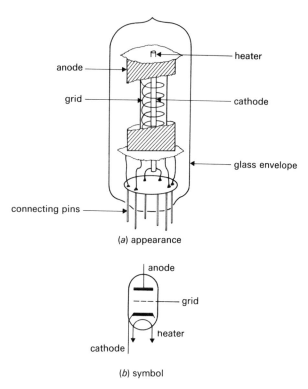

(*a*) appearance

(*b*) symbol

Figure 89   Triode valve and its circuit symbol.

anode. If the control grid is made positive with respect to the cathode more electrons are attracted from the cathode space charge and travel to the anode. By varying the positive potential on the grid the flow of electrons can be controlled. If the potential of the control grid is increased negatively with respect to the cathode, fewer electrons reach the anode, until eventually, when the negative potential of the control grid is just greater than the cathode potential, no electrons reach the anode. Consequently the anode current can be controlled by variation of the control grid potential.

Unlike the diode valve it is not possible to represent the behaviour of the triode valve by one characteristic since the anode current is controlled not only by the anode potential but also by the control grid potential. Thus two sets of static characteristics are plotted; one is of anode current against anode voltage for various fixed values of control grid potential; and the other of anode current against control grid potential for various fixed values of anode potential.

Solutions to self-assessment questions (pages 74–6)

1   Negative space charge.

2   Thermionic emission.

3   See Figure 87 (i) and (ii).

4   (a) The operating temperature.
    (b) The material used.

5   Directly heating.

6   Indirectly heating.

7 | A | 6 |
    |---|---|
    | B | 5 |
    | C | 4 |
    | D | 2 |
    | E | 7 |
    | F | 5 |
    | G | 3 |

8   Saturation.

9   Negative   positive.

10 TRUE.

11 See Figure 88.

12 (a) anode voltage
    (b) cathode temperature.

13 The saturation voltage is 100 V.

14 (b) Increased.

15 (i) thermionic diode: a, b, d, e, g, j
    (ii) semiconductor diode: a, c, f, g, h, k, l, n, p

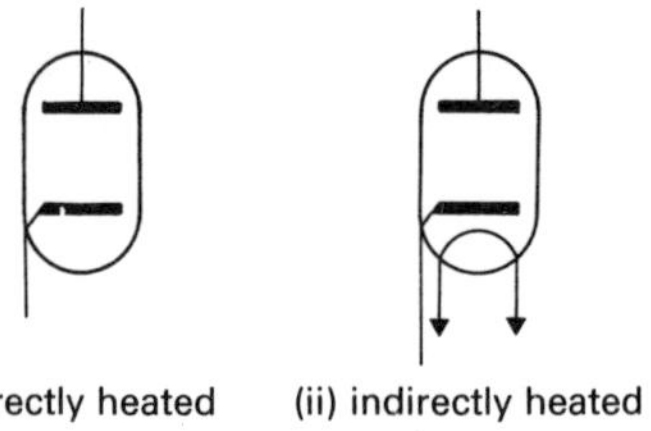

(i) directly heated   (ii) indirectly heated

Figure 87   Circuit symbols for diode valves.

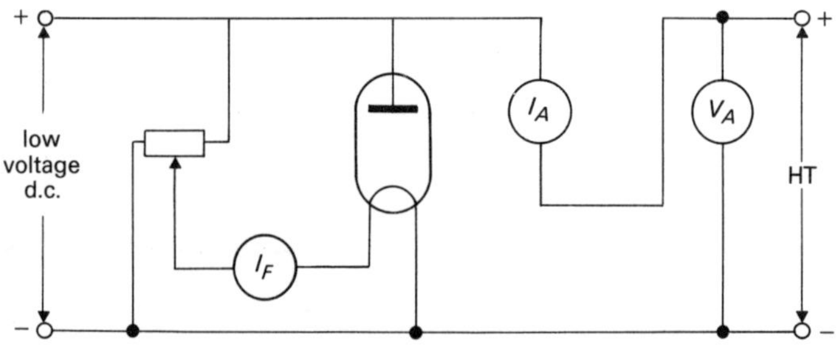

Figure 88   Solution for Question 11.

After reading the following material, the reader shall:

10.5 Sketch a circuit diagram for determining the static characteristics of the thermionic triode valve.

10.6 State the method for obtaining the static characteristics of the thermionic triode valve.

10.7 Plot and describe typical families of curves of I_A/V_A (output characteristics) and I_A/V_G (transfer characteristics).

Figure 90 shows a typical circuit diagram for determining the static characteristics of a Mullard 6J5 triode valve.

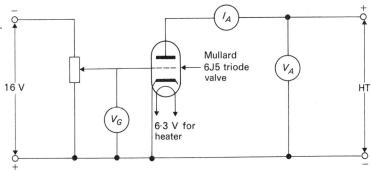

Figure 90 Circuit for determining the static characteristics of a triode valve.

(i) *I_A/V_A characteristics*

The control grid voltage, V_G, was adjusted to 0.8 V and maintained constant. The anode voltage, V_A, was increased in suitable increments from zero to about 300 V (or until I_A reached 15 mA). Values of anode voltage and anode current were recorded at each increment. The values recorded are shown in Table 8. The procedure was repeated for grid voltages of -6 V, -4 V, -2 V and 0 V.

Table 8 I_A/V_A characteristics: typical results obtained for a Mullard 6J5 triode valve.

$V_G = -8$ V		$V_G = -6$ V		$V_G = -4$ V		$V_G = -2$ V		$V_G = 0$ V	
V_A(Volts)	I_A(mA)	V_A(Volts)	I_A(mA)	V_A(Volts)	I_A(mA)	V_A(Volts)	I_A(mA)	V_A(Volts)	I_A(mA)
30	0	30	0	30	0	30	0·12	10	1·05
60	0	60	0	60	0·05	60	2·2	30	2·82
90	0	90	0	90	1·15	90	4·9	50	4·95
120	0	120	0·7	120	3·7	120	8·4	70	7·25
150	0·4	150	2·65	150	7	150	11·8	90	9·75
180	1·85	180	5·6	180	11	168	15	110	13
210	4·45	210	9	211	15	—	—	126	15
240	7·3	240	13	—	—	—	—	—	—
270	11·1	270	15	—	—	—	—	—	—
300	15	300	—	—	—	—	—	—	—

(ii) I_A/V_G *characteristics*

The anode voltage, V_A, was adjusted to 250 V and maintained constant. The control grid voltage, V_G, was varied in suitable increments from -15 V to zero until the anode current (I_A) reached 15 mA. Values of anode current and grid voltage were recorded at each increment as shown in Table 9. The procedure was repeated for constant anode voltages of 200 V and 150 V.

From the results of Tables 8 and 9 graphs of I_A/V_A and I_A/V_G were plotted as shown in Figures 91 and 92.

Table 9 I_A/V_G characteristics: typical results obtained for a Mullard 6J5 triode valve.

$V_A = 150$ V		$V_A = 200$ V		$V_A = 250$ V	
V_G(V)	I_A(mA)	V_G(V)	I_A(mA)	V_G(V)	I_A(mA)
-15	0	-15	0	-15	0
-14	0	-14	0	-14	0·2
-13	0	-13	0	-13	0·65
-12	0	-12	0	-12	1·25
-11	0	-11	0	-11	2·4
-10	0	-10	0·2	-10	3·9
-9	0	-9	0·75	-9	5·9
-8	0·35	-8	1·7	-8	8·25
-7	1·1	-7	3·25	-7	10·5
-6	2·5	-6	5·3	-6	13·5
-5	4·5	-5	7·55	$-5·5$	15
-4	7	-4	10	—	—
-3	9·6	-3	13	—	—
-2	12	-2	15	—	—
-1	15	—	—	—	—

(i) I_A/V_A *characteristics* (*output characteristics*)

Figure 91 shows the anode current, I_A, against anode voltage, V_A, characteristic for a Mullard 6J5 triode valve. It can be seen that the characteristics are approximately linear, except at low values of anode current, and are equally spaced for equal increments of control

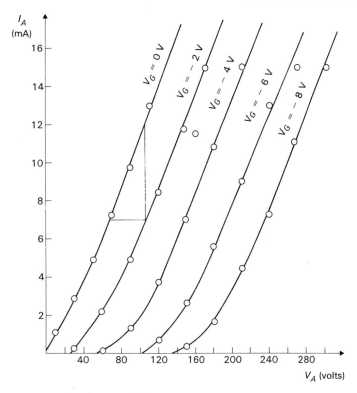

Figure 91 I_A/V_A characteristics.

grid potential. This means that the change in anode current is proportional to the change in anode potential for constant control grid potential. When the control grid potential is zero volts the triode is behaving as a diode. Also anode current flows only if the anode potential is made sufficiently positive to overcome the negative potential of the control grid.

(ii) I_A/V_G characteristics (*transfer characteristics*)

Figure 92 shows the anode current, I_A, against control grid voltage, V_G, characteristic for a Mullard 6J5 triode valve. It can be seen that the characteristics are again practically linear and are equally spaced for equal increments of anode voltage. This means that the change of anode current is proportional to the change in grid potential for constant anode potential. For each characteristic as the control grid potential is increased negatively, the anode current decreases until the cut off point is reached and no anode current flows, i.e. $I_A = 0\,\text{V}$.

It is clear from the preceding comments that the anode current is more sensitive to changes in control grid voltage than changes in anode voltage.

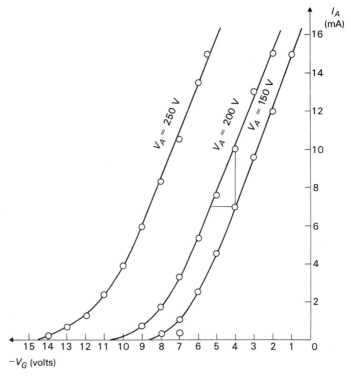

Figure 92 I_A/V_G characteristics.

After reading the following material, the reader shall:

10.8 Define anode slope resistance (r_a), mutual conductance (g_m) and amplification factor (μ).

10.9 State the relationship between r_a, g_m and μ.

10.10 Determine the value of r_a, g_m and μ from given characteristics.

The behaviour of a triode valve can be expressed in terms of three constants, all of which may be obtained from the static characteristics.

(i)   *Anode slope resistance, r_a*

The BS preferred term is the anode a.c. resistance and is defined as the ratio of the change in anode voltage (ΔV_A) to the change in anode current (ΔI_A) for constant control grid potential (V_G).

i.e. anode a.c. resistance, $r_a = \dfrac{\Delta V_A}{\Delta I_A}$ for constant V_G

r_a is usually expressed in kΩ since I_A is measured in mA.

The anode slope resistance may be obtained directly from the linear portion of the I_A/V_A characteristic of Figure 91.

Anode a.c. resistance

$$= \frac{1}{\text{slope of } I_A/V_A \text{ characteristic for constant } V_G}$$

$$= \frac{V_A}{I_A}$$

Using the graphs of Figure 91,

At $V_G = 0\,\text{V}$, anode a.c. resistance $= \dfrac{120 - 60}{14 - 6} \dfrac{\text{volts}}{\text{mA}}$

$$= \frac{60}{8}\,\text{k}\Omega$$

$$= 7 \cdot 5\,\text{k}\Omega$$

At $V_G = -8\,\text{V}$, anode a.c. resistance $= \dfrac{300 - 244}{15 - 8}$

$$= \frac{56}{8}$$

$$= 7\,\text{k}\Omega$$

(ii)   *Mutual conductance g_m*

This is defined as the ratio of change in anode current (ΔI_A) to the change in grid voltage (ΔV_G) for constant anode voltage (V_A).

$$\text{Mutual conductance, } g_m = \frac{\Delta I_A}{\Delta V_G} \text{ for constant } V_A$$

g_m is usually expressed in mA/volt.

Mutual conductance may be obtained directly from the linear portion of the I_A/V_G characteristic of Figure 92.

Mutual conductance = slope of I_A/V_G characteristic for constant V_A

$$= \frac{I_A}{V_G}\,\text{mA/volt}$$

Using the graphs of Figure 92,

At $V_A = 150\,\text{V}$, $g_m = \dfrac{15 \cdot 8 - 6}{9 - 5} \dfrac{\text{mA}}{\text{V}}$

$$= \frac{9 \cdot 8}{4}\,\text{mA/V}$$

$$= 2 \cdot 45\,\text{mA/V}$$

$$\text{At } V_A = 250\,\text{V}, g_m = \frac{15-6}{4\cdot4-1}\frac{\text{mA}}{\text{V}}$$

$$= \frac{9}{3\cdot4}\,\text{mA/V}$$

$$= 2\cdot64\,\text{mA/V}$$

(*iii*)   *Amplification factor, μ*

This is a measure of the relative effectiveness of the control grid and anode in controlling the anode current. Amplification factor is defined as the change of anode voltage (ΔV_A) to the change in grid voltage (ΔV_G) for the *same change* in anode current.

$$\text{Amplification factor, } \mu = \frac{\Delta V_A}{\Delta V_G} \text{ for constant } I_A$$

$$= \frac{\Delta V_A}{\Delta I_A} \times \frac{\Delta I_A}{\Delta V_G}$$

i.e. $\underline{\underline{\mu = r_a \times g_m}}$

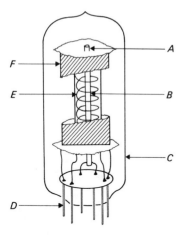

Figure 93   Triode valve.

Self-assessment questions

16   Figure 93 shows the constructional features of a thermionic triode valve. Match the component parts labelled *A* to *F* to the numbers 1 to 6.

A	1 connecting pins
B	2 anode
C	3 grid
D	4 glass envelope
E	5 cathode
F	6 heater

17   Complete the following statement:
The triode valve has three electrodes. These are:

(i)   ——————
(ii)   ——————
(iii) ——————

18   Statement 1: The control grid, in the form of a wire mesh is sited between the anode and cathode.

Statement 2: The purpose of the control grid is to control the flow of electrons from cathode to anode.

(*a*) Only statement 1 is true.
(*b*) Only statement 2 is true.
(*c*) Both statements 1 and 2 are true.
(*d*) Neither statement 1 nor 2 is true.

Underline the correct answer.

19 For anode current to flow the control grid is made negative with respect to the cathode.

TRUE/FALSE

20 For no anode current to flow the control grid is made more negative with respect to the cathode than the positive potential of the anode.

TRUE/FALSE

21 Draw the circuit symbol of a triode valve.

22 Sketch a typical anode current/anode voltage characteristic of a triode valve for constant values of grid voltage. Indicate on the characteristic the highest grid potential.

23 Sketch a typical anode current/grid voltage characteristic of a triode valve for constant values of anode voltage. Indicate on the characteristic the highest anode potential.

Complete statements 9 to 12.

24 The anode a.c. resistance is defined as $r_a = \dfrac{\Delta\underline{\quad}}{\Delta\underline{\quad}}$ for constant

_____.

25 The mutual conductance is defined as $g_m = \dfrac{\Delta\underline{\quad}}{\Delta\underline{\quad}}$ for constant

_____.

26 The amplification factor is defined as $\mu = \dfrac{\Delta\underline{\quad}}{\Delta\underline{\quad}}$ for constant

_____.

27 The relationship between r_a, g_m and μ is

$\mu =$

Solutions to self-assessment questions (pages 84 and 85)

16 A   6
    B   5
    C   4
    D   1
    E   3
    F   2

17
(i) Anode
(ii) Cathode
(iii) Control grid

18 (c)

19 FALSE. For anode current to flow the control grid is made positive with respect to the cathode.

20 TRUE.

21 See Figure 94.

22 See Figure 95.

23 See Figure 96.

24 $r_a = \dfrac{\Delta V_A}{\Delta I_A}$ for constant V_G

25 $g_m = \dfrac{\Delta I_A}{\Delta V_G}$ for constant V_A

26 $\mu = \dfrac{\Delta V_A}{\Delta V_G}$ for constant I_A

27 $\mu = r_a \times g_m$

Figure 94   Solution for Question 21.

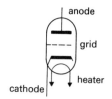

Figure 95   Solution for Question 22.

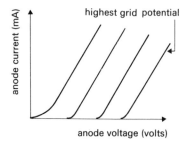

Figure 96   Solution for Question 23.

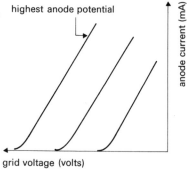

Topic area Cathode ray tubes

After reading the following material, the reader shall:

10   Know the principles of operation of a cathode ray tube (CRT).
10.1   Label a diagram of a CRT.
10.2   State the operation of a CRT.
10.3   State the function of each part of the CRT.
10.4   State that deflection can be produced by electric and/or magnetic fields.
10.5   Distinguish between electric and magnetic field deflection.

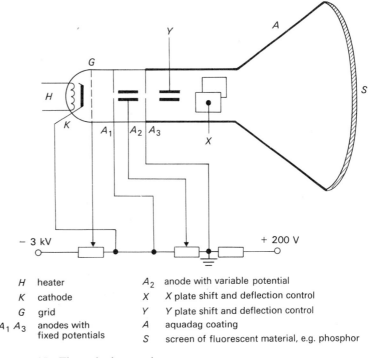

H	heater	A_2	anode with variable potential
K	cathode	X	X plate shift and deflection control
G	grid	Y	Y plate shift and deflection control
$A_1 A_3$	anodes with fixed potentials	A	aquadag coating
		S	screen of fluorescent material, e.g. phosphor

Figure 97   The cathode ray tube.

Figure 97 shows a labelled diagram of a typical cathode ray tube and its associated circuitry. It consists of three basic units

(i)   the *electron gun*
(ii)   the *deflection system*
(iii)   the *fluorescent screen*

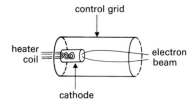

Figure 98   The electron gun.

The electron gun, shown in Figure 98, is the structure and arrangement of the electrodes. It consists of three basic elements, the heater, the cathode and the control grid. The heater is similar in structure and function to that in a thermionic valve. The cathode is a short nickel cylinder which is oxide coated and closed at one end. It is heated by the filament; thermionic emission takes place and electrons are liberated. The cathode structure is such that the stream of electrons liberated flows entirely forwards towards the screen. The control grid is cylindrical, being open at the cathode end and closed at the other end except for a very small aperture. The control grid functions in a similar way to the grid of a thermionic triode valve. It is always of negative potential with respect to the cathode. Consequently the function of the control grid is to shape the stream of electrons into a beam and control the intensity of the electron beam.

Anodes A_1 and A_3 are accelerating electrodes. Their function is to accelerate the electron beam towards the screen (S) by increasing the velocity of the beam. These two anodes are operated at a fixed potential.

Anode A_2 is the focusing anode and its potential may be varied. This enables the electron beam to be focussed to a fine spot on the screen of the cathode ray tube.

In order for the spot to trace voltage and current waveforms on the screen it is necessary to move the spot on the screen. This is the function of the X and Y deflection system.

Conversion of the beam of electrons into a visual picture is achieved by the bombardment on the fluorescent screen of the electron beam. Consequently the purpose of the screen coating is twofold: *luminescence* to convert the electron beam into light, and *phosphorescence* to continue to glow after the initial impact of the beam. Zinc orthosilicate is a typical screen material.

An aquadag coating (A) on the inner surface of the glass envelope connects the screen electrically to anode A_3. This provides a return path to the cathode via the power supply for the electrons after bombardment of the screen.

The electron gun is designed to produce a spot on the cathode ray tube screen. In order that this spot can indicate various types of current and voltage on the screen it is necessary to move the spot in two directions and trace out the particular waveform on the screen. This is the function of the deflection system; the electron beam is deflected in two directions at right angles to each other, one direction vertically and the other horizontally. Deflection is achieved by using either an electric field or a magnetic field.

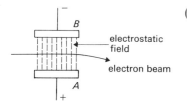

Figure 99 Electric field deflection.

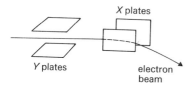

Figure 100 Two sets of plates for electrostatic deflection.

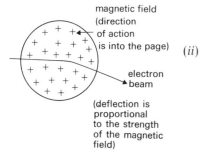

Figure 101 Magnetic field deflection.

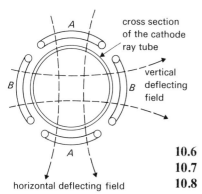

A horizontal deflecting coil
B vertical deflecting coil

Figure 102 Two sets of coils for magnetic deflection.

(i) *Electric field deflection*

The principle of electrostatic deflection is shown in Figure 99. If plate *A* is made positive with respect to plate *B* the electrons are repelled from plate *B* and attracted towards plate *A* (since like charges repel and unlike charges attract). The amount of deflection, and hence movement of the spot, is proportional to the voltage applied to the plates. If plate *B* is now made positive with respect to plate *A*, electrons are repelled by plate *A* and attracted towards plate *B*. Since the electron beam needs to be deflected in two directions, two sets of plates are required at right angles to each other as shown in Figure 100.

The first set of plates deflect the beam vertically and are known as the *Y* deflection plates (Figure 100). The second set of plates deflect the beam horizontally and are known as the *X* deflection plates (Figure 100). Consequently by variation of the potential on the two pairs of plates the electron beam can be moved to any part of the screen.

This method of deflection of the electron beam is used for cathode ray tubes which are employed in oscilloscopes.

(ii) *Magnetic field deflection*

Magnetic field deflection operates in a similar manner to the electrostatic deflection system except that the electron beam is deflected by means of a magnetic field, the amount of deflection being proportional to the strength of the magnetic field, as shown in Figure 101.

The beam is again deflected in two directions at right angles by two sets of coils as shown in Figure 102.

This method of deflection is used in television tubes to obtain wide angled deflection; the two sets of coils being placed at the same position along the length of the tube.

After reading the following material, the reader shall:

10.6   State the purpose of a time base in an oscilloscope.
10.7   Identify the time base signal as a sawtooth waveform.
10.8   Identify the scan and flyback sections of the time base waveform.

If an alternating signal is applied to the *Y* plates of an oscilloscope without any potential on the *X* plates, then only a vertical trace is seen on the screen. However, if a voltage is applied to the *X* plates

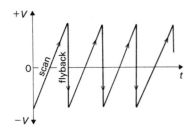

Figure 103 The time base sawtooth waveform.

simultaneously with the a.c. signal applied to the *Y* plates, the correct waveform is observed on the screen. The effect of this voltage applied to the *X* plates is to move the spot horizontally at constant speed. This is achieved by applying a sawtooth waveform to the *X* plates as shown in Figure 103.

The sawtooth consists of two sections, the *scan* and the *flyback*. The scan part of the sawtooth wave moves the spot horizontally at constant speed across the screen. The flyback part of the sawtooth wave follows at the end of the scan when the voltage is reduced to zero. The scan frequency is usually an exact multiple or sub multiple of the frequency of the signal being examined. The sawtooth waveform applied to the *X* plates is commonly known as the time base signal and is variable, so that signals of any frequency can be displayed on the screen.

After reading the following material, the reader shall:

10.9 State the functions of the following operating controls normally found on an oscilloscope:

(i) brilliance (brightness)
(ii) focus
(iii) *Y* controls
(iv) *X* controls
(v) time base
(vi) stability and trig level
(vii) a.c./d.c. switch

10.10 Follow a procedure to set up and use the oscilloscope and its associated controls to display any waveform.

Figure 104 shows the front panel of a Telequipment Oscilloscope S51B with the operating controls.

The brilliance or brightness control controls the brightness of the spot by adjusting the grid-cathode potential. It is important that high levels of brilliance should not be used, since they permanently damage the fluorescent coating of the screen.

The focus control enables the spot diameter to be adjusted by variation of the potential of anode A_2 with respect to the fixed potentials of anodes A_1 and A_3 (see Figure 97).

The *Y* control consists of two controls on the same spindle. One control is calibrated in volts/cm (ranging from 0·1 to 50 V/cm) so that the oscilloscope can display signal amplitudes which may range from

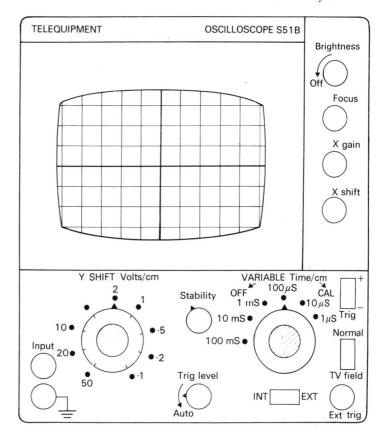

Figure 104   The front panel of a Telequipment S51B Oscilloscope.

millivolts to hundreds of volts. The other control is the *Y* shift control so that the signal trace may be displayed at any vertical position on the screen.

The *X* control also consists of two controls, one the *X* gain control which expands the signal trace about the centre of the screen, the other the *X* shift control which moves the signal trace horizontal.

The time base control consists of two controls on the same spindle; one a variable control which ranges from off to calibrate, and the other being calibrated in time/cm ranging from 100 ms/cm to 1 μs/cm. When making measurements using the time base it is important to turn the variable control to calibrate. The stability control enables a stable signal trace to be obtained and is used in conjunction with the trig level control which enables the starting point of the sweep to be selected from any point on the positive going slope of the displayed waveform. The trig level control operates in this manner only when switched from the 'auto' position.

The a.c./d.c. switch is normally used in the a.c. position. The switch is used in the d.c. position only when a d.c. level or very low frequency signal needs to be displayed or measured.

Setting up the oscilloscope before use

Before switching on the oscilloscope:

(*a*)   Set the *X* gain control fully anticlockwise (minimum position).
(*b*)   Set the *X* shift control to the mid position.
(*c*)   Set the volts/cm control to a suitable scale for the signal being displayed.
(*d*)   Set the *Y* shift control to the mid position.
(*e*)   Set the stability control fully anticlockwise.
(*f*)   Set the trig level control to the 'auto' position (fully anticlockwise).
(*g*)   Set the focus control to the mid position.
(*h*)   Set the brightness control fully anticlockwise (off position).
(*j*)   Select the a.c. position of the a.c./d.c. switch.
(*k*)   Set the variable time base control fully anticlockwise to the calibrate position.
(*l*)   Set the time/cm control to a suitable scale for the signal being displayed.

Setting up the oscilloscope after switching on

(*a*)   Switch on the oscilloscope and allow a few minutes for the cathode ray tube to attain operating temperature.
(*b*)   Apply the signal to the input terminals.
(*c*)   Adjust the brightness and focus controls to obtain a sharp trace on the screen.
(*d*)   Centralize the trace using the *X* and *Y* shift controls.
(*e*)   Rotate the stability control anticlockwise until the trace is stable on the screen.

After reading the following material, the reader shall:

10.11   Interpret the data on the oscilloscope to measure the amplitude of various waveforms.
10.12   Interpret the data on the oscilloscope to measure the frequency of various waveforms.

Having adjusted the controls of the oscilloscope (as in the foregoing objectives) to obtain a stable trace of the signal under test, its amplitude and frequency can now be determined.

The amplitude is determined directly from the graticule scale of the oscilloscope. This is scaled in centimetre squares. Consequently the amplitude of any signal can be measured in centimetres directly from the graticule. The amplitude in volts can now be found by multiplying the amplitude length in centimetres by the setting of the volts/cm scale:

$$\text{Amplitude (volts)} = \text{length (cm)} \times \text{volts/cm}$$

The frequency of the waveform can be calculated using the time base control. First set the variable time base control to 'calibrate' and set the time base control in time/cm to a suitable position for a convenient trace. Measure in centimetres horizontally the distance required to complete a whole number of cycles. The frequency is then given by:

$$\text{Frequency (Hz)} = \frac{\text{number of cycles}}{\text{time/cm} \times \text{number of cm}}$$

Figure 105 shows a typical waveform displayed on the oscilloscope. Only the relevant controls are shown for clarity in order to measure the magnitude and frequency.

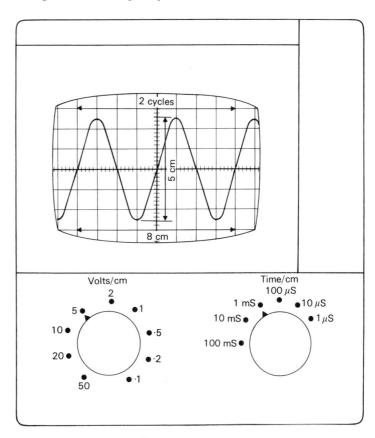

Figure 105 Typical waveform obtained on the oscilloscope.

$$\text{Amplitude (volts)} = \text{length (cms)} \times \text{volts/cm}$$
$$= 5 \times 5$$
$$= \underline{\underline{25 \text{ volts}}}$$

$$\text{Frequency (Hz)} = \frac{\text{number of cycles}}{\text{time/cm} \times \text{number of cm}}$$
$$= \frac{2}{1 \times 10^{-3} \times 8}$$
$$= \frac{2 \times 10^{3}}{8} = \underline{\underline{250 \text{ Hz}}}$$

Self-assessment questions

1 Match the component parts *A* to *H* with those marked 1 to 10 on the diagram of the cathode ray tube in Figure 106. Each letter may be used once, more than once or not at all.

1	
2	*A* screen
3	*B* heater
4	*C* anode
5	*D* X plate shift
6	*E* aquadag coating
7	*F* cathode
8	*G* Y plate shift
9	*H* grid
10	

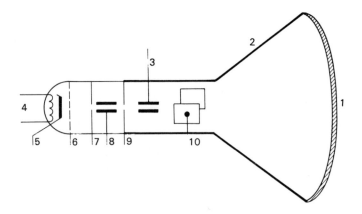

Figure 106 The cathode ray tube.

2 The main function of the electron gun is to:

(*a*) accelerate electrons
(*b*) produce electrons
(*c*) deflect electrons
(*d*) focus electrons

Tick the correct answer.

3 The main function of the two cylindrical anodes is to:

(*a*) accelerate electrons
(*b*) produce electrons
(*c*) deflect electrons
(*d*) focus electrons

Tick the correct answer.

4 The focusing of the electron beam is controlled by varying:

(*a*) the heater voltage
(*b*) the grid bias voltage
(*c*) the secondary accelerating voltage with respect to the first
(*d*) the first electrode voltage with respect to the cathode

Tick the correct answer.

5 The intensity of the electron beam is controlled by varying:

(*a*) the heater voltage
(*b*) the grid bias voltage
(*c*) the accelerating voltage
(*d*) the deflecting voltage

Tick the correct answer.

6 Complete the following.
The three main parts of the cathode ray tube are:

(*a*) _____
(*b*) _____
(*c*) _____

7 The focus control on a cathode ray tube enables the diameter of the spot to be adjusted.

TRUE/FALSE

8 The intensity control enables the operator to adjust the shape of the spot on the screen so that it remains circular.

TRUE/FALSE

9   In a cathode ray tube, deflection of the electron beam can be achieved either by electric field deflection or magnetic field deflection.

Statement 1: Electric field deflection is achieved by electrostatic forces between two parallel plates.

Statement 2: Magnetic field deflection is achieved by magnetic forces from an electromagnet or permanent magnet, which is external to the tube.

(*a*)   Only statement 1 is true.
(*b*)   Only statement 2 is true.
(*c*)   Both statements 1 and 2 are true.
(*d*)   Neither statements 1 nor 2 is true.

Underline the correct answer.

10   Complete the following:
The functions of the cathode ray tube screen coating are

(i)   _____
(ii)   _____

11   The time base control on a cathode ray oscilloscope is usually calibrated in time/min.

TRUE/FALSE

12   The spot on the screen of a cathode ray oscilloscope may be centred on the graticule using the X and Y shift controls with the time base control in the external position.

TRUE/FALSE

13   It is possible to display one complete cycle of an alternating signal providing the time base is switched to a time/cm setting.

TRUE/FALSE

14   The sawtooth time base signal consists of two sections. These are

(i)   _____
(ii)   _____

15   Figure 107 shows a square wave signal displayed on the screen of a cathode ray oscilloscope.
What is

(*a*)   the magnitude   _____ volts
(*b*)   the frequency   _____ Hz

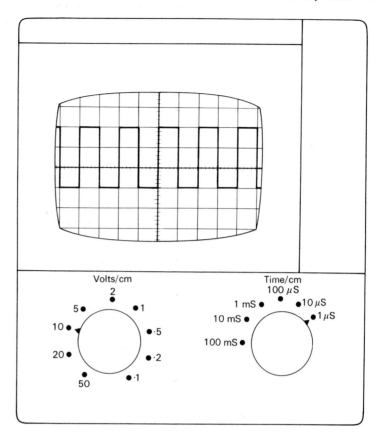

Figure 107 Oscilloscope waveform for Question 15.

Solutions to self-assessment questions (pages 94–97)

1   1   A
    2   E
    3   G
    4   B
    5   F
    6   H
    7   C
    8   C
    9   C
   10   D

2   (*b*) Produce electrons.

3   (*a*) Accelerate electrons.

4   (*c*) The secondary accelerating anode with respect to the first.

5   (*b*) The grid bias voltage.

6   (*a*) Electron gun.
    (*b*) Deflection system.
    (*c*) Fluorescent screen.

7   TRUE.

8   FALSE. The intensity control controls the brightness of the spot.

9   (*c*).

10   (i) Fluorescence
     (ii) Phosphorescence

11   FALSE. It is calibrated in time/cm.

12   TRUE.

13   TRUE.

14   (i) the scan
     (ii) the flyback

15   (*a*) Magnitude (volts) = length (cm) × volts/cm

$$= 3 \times 10$$

$$= \underline{30 \text{ volts}}$$

(*b*) Frequency (Hz) = $\dfrac{\text{number of cycles}}{\text{time/cm} \times \text{number of cm}}$

$$= \frac{5}{1 \times 10^{-6} \times 10}$$

$$= \frac{5 \times 10^{6}}{10}$$

$$= 500\,000$$

$$= \underline{0.5 \text{ MHz}}$$

Topic area Small signal amplifiers

After reading the following material the reader shall:

12 Know the principle of simple amplifier operation.
12.1 State the purpose of an amplifier.
12.2 State the difference between small signal and large signal amplifiers.
12.3 Draw the BS 3939 symbol for an amplifier.
12.4 State that an amplifier has the following properties:

(*a*) amplification or gain
(*b*) frequency response
(*c*) transfer characteristic

12.5 Explain the distortion occuring due to the frequency response and transfer characteristic of an amplifier.

The purpose of an amplifier is to increase the amplitude of the voltage or current or power of an electrical signal.

When an amplifier is used to amplify either the voltage or current of an input signal such that the output signal is not distorted it is said to be a *small signal* amplifer. The parameters of the device are assumed to remain constant during the period of operation of the device.

Parameter is a word now used to represent any term that aids description of how a device (or system) behaves within quoted boundaries. For instance, some of the parameters of resistance wire are length, diameter, and resistivity.

When an amplifier is used to amplify the power of an input signal it is said to be a *large signal* amplifier. The parameters of the device are not constant during the period of operation of the device.

A simple example of this is the audio power amplifier where maximum power at audio frequencies is transferred from a small signal voltage amplifier to a loudspeaker driven by a power amplifier.

In the diagram shown in Figure 108 the amplifier can be drawn within the box or by itself. The 'blunt' end represents the input to the amplifier and the 'sharp' end represents the output of the amplifier. Although the two connection symbol is not a standard symbol it will be used where it aids the explanation of the system under discussion.

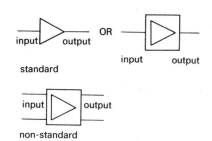

Figure 108 BS 3939 symbol for an amplifier.

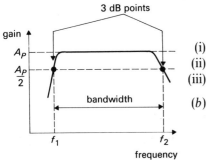

Figure 109 Frequency response of an amplifier.

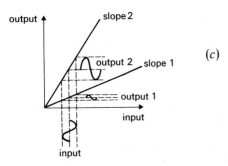

Figure 110 Theoretical transfer characteristic.

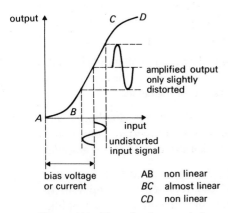

Figure 111 Transfer characteristic of a typical amplifying device.

(*a*) The *gain* of an amplifier is the amount of increase of the output signal amplitude compared with the amplitude of the signal before amplification.

The amount of amplification depends on:

(i) the number of amplifier stages used
(ii) the design of the amplifier stage
(iii) the characteristics of the device used in the amplifier stage

(*b*) If the frequency of an input signal (of constant amplitude) to an amplifier is varied and the gain recorded, a graph can be plotted of gain against frequency. The most convenient scales for the axis are chosen as required, the gain being measured as a ratio or in logarithmic units of gain. The way in which the gain varies with frequency gives an indication of the *frequency response* of an amplifier. A typical graph is shown in Figure 109.

When the power gain falls to half the constant or mid-band value (another way of saying this which will be dealt with in *Electronics* 3 is where the gain falls by 3 dB) a practical indication of the *bandwidth* or *passband* of an amplifier is obtained, i.e. $f_2 - f_1$.

(*c*) The *transfer characteristic* of an amplifier indicates the amount of gain that may be obtained from an amplifying device. This can be seen from the graph of output signal against input signal shown in Figure 110.

It can be seen that the steeper the slope, the greater the amplification. In practice the slope is *non linear* and hence the output can be severely distorted unless careful design is used, as shown in Figure 111.

By positioning the *mean level* of the input signal on the most linear part of the curve the amount of distortion produced is minimized.

The position of the input signal mean level on the most linear part of the curve is achieved by biasing, about which more information will be given in a later section in this topic area.

From the diagram shown in Figure 109 the amplifier can be considered to be a filter that passes only a band of approximately constant amplitude signals between frequencies $f_2 - f_1$. Outside this band of frequencies the amplitude of the signals passed *decreases* rapidly. Hence an input signal that contains frequencies outside this pass band cannot be reproduced exactly at the output of the amplifier because signals outside the pass band are amplified at values different from those inside the pass band. This causes distortion of the output signal, known as *frequency distortion*. This distortion can be reduced using feedback as shown in topic area Feedback of *Electronics* 3.

Consider the diagram shown in Figure 112 where the input signal is applied at the point shown. The output is grossly distorted because

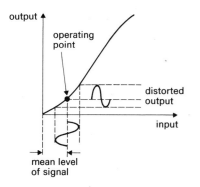

Figure 112 Distortion using transfer characteristic.

the operating point is not on the most linear part of the curve. This causes distortion of the output signal known as *amplitude distortion*.

It can also be seen that if the operating point is at the centre part of the curve and the input signal has too large an amplitude, amplitude distortion occurs. This distortion can be minimized by choosing the most linear part of the characteristic, and by restricting the amplitude of the input signal to the lowest acceptable value.

Self-assessment questions

Complete the following sentences either by writing in the correct word or by crossing out the incorrect word where applicable. The questions all relate to small signal amplifiers except where otherwise stated.

1 The purpose of a small signal amplifier is to increase the amplitude/ frequency of the voltage of an electric signal.

2 Large signal amplifiers are used to amplify the ＿＿＿＿＿＿ of an electrical signal.

3 The gain of an amplifier is the amount of increase of the output signal amplitude compared with the ＿＿＿＿＿＿ of the signal before amplification.

4 A graph of voltage/gain plotted against frequency gives a practical indication of the frequency response of the amplifier.

5 The graph which gives an indication of the amount of gain that may be obtained from an amplifying device is called the characteristic. The steeper/flatter the slope of this graph the greater is the amplification of the device.

6 To obtain minimum amplitude distortion in practice the mean level of the input signal is placed on the part of the curve that is most ＿＿＿＿＿＿. The positioning on the curve is achieved by what is known as ＿＿＿＿＿.

7 Write down two factors that determine the amount of amplification of an amplifier.

8 Explain the term frequency distortion.

After reading the following material, the reader shall:

13 Know for the small signal common emitter amplifier:

(a)   the circuit
(b)   the operation
(c)   the construction of a d.c. load line on the transistor characteristics
(d)   the use of a d.c. load line on the transistor characteristics

13.1 Explain the difference between the static and dynamic transfer characteristic of the transistor in the common emitter mode.

13.2 Draw the circuit diagram of a single stage amplifier having a load resistor R_L.

13.3 Show that the supply voltage $V_{CC} = I_C R_L + V_{CE}$.

13.4 Explain that bias is required to give a selected quiescent operating point on the output characteristic.

13.5 Explain the load line construction on a given set of output characteristics for a common emitter amplifier.

13.6 Explain the effect of a small sinusoidal current input on the quiescent condition.

13.7 State that voltage inversion occurs between input and output signals.

For this explanation the transistor is assumed to be of the n–p–n type. (A similar explanation can be made taking account of the correctly applied potentials for the p–n–p type.) From the topic area Elementary theory of semiconductors it can be seen that the transistor is a current operated device and the transfer characteristic with no load resistor in the collector circuit is called the *static transfer characteristic* as shown in Figure 113, i.e. $R_L = 0$.

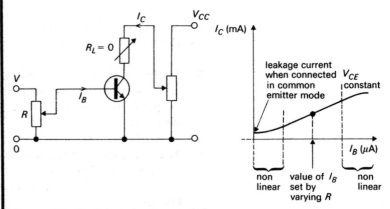

Figure 113   Static transfer characteristic.

From this graph, provided that I_B does not drop to less than approximately 10 μA, linearity is achieved. A reasonable value of I_B

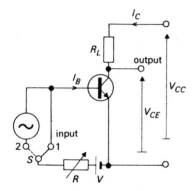

Figure 114 Common emitter circuit where R_L does not equal zero.

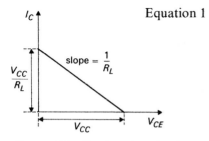

Figure 115 Graph of Equation 1.

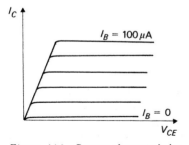

Figure 116 Output characteristics.

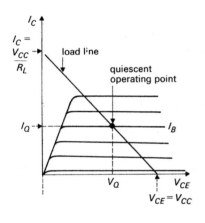

can be set by varying R to ensure that the transistor is operated at the mid-point of the linear part of the curve. The way in which this value of bias current is set in a practical circuit is considered in a later section in this topic area.

What effect does the increasing of R_L have on the shape of the static characteristics? Consider Figure 114 where R_L has been set to a particular value.

A graph can be plotted of I_C against I_B for the particular value of load resistance. A different plot will be obtained for each different value of load resistance and each individual plot is the *dynamic transfer characteristic* for that value of load. This can be measured in practice, or the following method can be used.

Equation 1 If Kirchhoff's laws are applied to the collector emitter circuit then the applied voltage $V_{CC} = V_{CE} + I_C R_L$.

Now V_{CC} is a constant value
$\qquad R_L$ is a constant value

Thus Equation 1 represents the way in which I_C varies as V_{CE} varies for a particular load resistance R_L and thus can be plotted on a graph.

Rearranging Equation 1 gives

$$I_C = -\frac{1}{R_L}\cdot V_{CE} + \frac{V_{CC}}{R_L}$$

If I_C is plotted against V_{CE} a straight line graph is obtained with slope $-(1/R_L)$. This graph is called a *load line*.

If I_B is set to make $I_C = 0$, then $V_{CE} = V_{CC}$.

If I_B is set such that the transistor is driven hard on and $V_{CE} = 0$ (which is not strictly true in practice) then

$$\frac{V_{CC}}{R_L} = I_C$$

This gives two points of the load line which is plotted as shown in Figure 115.

However, the output characteristics also show how the collector current I_C varies as the base current I_B varies, as shown in Figure 116.

When the two graphs are superimposed the load line intersects with the output characteristics. For a particular value of I_B this gives the point Q called the *quiescent operating point*. As can be seen from Figure 117 this gives the current I_Q flowing in the collector circuit and the voltage V_Q across the transistor for a base current of I_B.

Figure 117 Combined graphs.

The dynamic transfer characteristic is found by noting each value of I_C where the load line crosses the output characteristics for each value of I_B and then plotting I_C against I_B as shown in Figure 118. This has been shown for two output characteristics to show how the transfer characteristic varies in linearity for a theoretical and a practical plot of output characteristics.

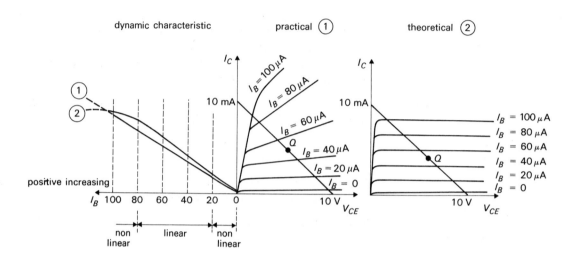

Figure 118   Plotting dynamic characteristic from graphs shown.

Providing the output curves are reasonably parallel and equidistance apart for equal increments of I_B then the dynamic transfer characteristic is reasonably linear. The value of I_B can be set so as to ensure that the transistor is operated on the most linear part of the dynamic transfer curve. From Figure 118 an approximate method of obtaining this point Q is to arrange for I_B to fix the point approximately half way along the load line.

Thus once R_L is fixed a dynamic transfer characteristic of the device can be obtained and this can be plotted

(i)   from measurements taken

or

(ii)   the knowledge of the static output characteristics and the load line.

There is only one such characteristic for each different value of load.

What now happens if the switch in Figure 114 is changed to position 2 and a small magnitude sinusoidal voltage generator is applied to the transistor base?

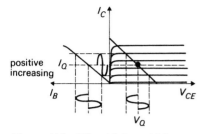

Figure 119 Effect of sinusoidal current input.

The base current varies sinusoidally about the steady bias value as shown in Figure 119.

If the base current increases then the collector current increases. As R_L is constant, from Ohm's law the voltage across the resistor must increase. But the applied voltage V_{CC} is constant. Therefore the voltage across the transistor must *decrease* as the base current increases. From Figure 119 it can be seen from the load line, which shows how the collector current varies, that the voltage across the transistor is the inverse of the applied waveform. (Remember I_B, although drawn in the negative direction, is $+I_B$.)

Consider a diagram of these waveforms, as shown in Figure 120.

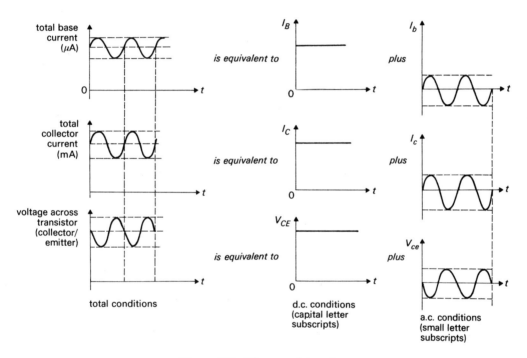

Figure 120 Diagram of waveshapes.

Each waveform is a *unidirectional* varying d.c. waveform. Each is equivalent to the waveform which could be made by combining a d.c. waveform together with an a.c. waveform.

In practice when the d.c. waveform is mentioned this is the d.c. waveform shown.

In practice when the a.c. waveform is mentioned this is the a.c. waveform shown.

It should be remembered that when a sinusoidal signal is applied at the base of the transistor a total waveform exists in the circuit.

From the diagram the output signal is the *inverse* of the input signal. Is this the same as a 180° phase shift of signal? Consider both the sinusoidal and triangular waveform shown in Figure 121.

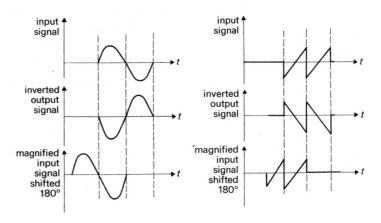

Figure 121   Signal inversion.

For the sinewave, the output signal has only half a wave shape similarity to the phase shifted wave.

For the triangular wave it can be even more clearly seen that the *inverted* output signal is *not* the same as the 180° phase shifted wave.

If the input circuit is purely resistive, then the voltage input signal has the same shape as the current waveform (different magnitude, same phase) and hence from the diagram shown in Figure 119 it can be seen that *voltage inversion* occurs between input and output signals.

Self-assessment questions

9   Using the circuit diagram shown in Figure 122 explain briefly the difference between the static transfer characteristic of the transistor and the dynamic transfer characteristic of the circuit.

Complete the following sentences which relate to Figure 122.

(i)   The position of the input signal mean level on the most linear part of the dynamic transfer characteristic is achieved by _____.

(ii) An increase in the base current causes the collector current to _____.

(iii) For a fixed value of load resistance, an increase in base current causes the collector–emitter voltage to _____.

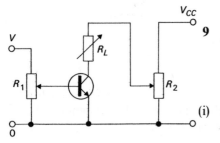

Figure 122   Circuit diagram for Question 9.

10

(i) Draw the circuit diagram of a single stage amplifier having a load esistor R_L.

(ii) From the circuit diagram show that the supply voltage

$$V_{CC} = I_C R_L + V_{CE}.$$

11

(i) When a small magnitude sinusoidal voltage generator is applied to the input of a common emitter amplifier, the output voltage is shifted through 180° with reference to the input voltage.

TRUE/FALSE

(ii) When the base current of a common emitter amplifier is increased by a small positive amount the collector–emitter voltage of the transistor increases.

TRUE/FALSE

12 Tick the correct alternative.

(i) A common emitter amplifier has the input and output signals at the following connections:

	input signal		output signal
A	base	and	emitter
B	base	and	collector
C	emitter	and	collector
D	emitter	and	base

(ii) If the load resistor R_L of a simple common emitter amplifier is increased in value, the source voltage V_{CC} remaining constant, then considering the load line construction on the output characteristics:

A The d.c. load line slope remains the same and moves away from the origin of the characteristics.

B The d.c. load line slope remains the same and moves towards the origin of the characteristics.

C The d.c. load line slope is increased.

D The d.c. load line slope is reduced.

Solutions to self-assessment questions (pages 106 and 107)

9   The transfer characteristic of a device indicates how the device affects the input signal and hence produces an output signal.

For the transistor only, when $R_L = 0$, the transfer characteristic is obtained by plotting a graph of I_C against I_B. This is called the static transfer characteristic.

When R_L has a fixed magnitude the transistor has become part of an amplifier. The transfer characteristic for the amplifier is obtained by plotting a graph of I_C against I_B and this is called the dynamic transfer characteristic.

(i)   Biasing
(ii)   Increase
(iii)   Decrease

10
(i)   See Figure 114, page 103.

(ii)   If Kirchoff's laws are applied to the collector emitter circuit then the applied voltage
V_{CC} = voltage across resistor + voltage across transistor
$\therefore V_{CC} = I_C R_L + V_{CE}$

11
(i)   FALSE. The output voltage is inverted.
(ii)   FALSE. The collector emitter voltage of the transistor decreases.

12
(i)   *B*
(ii)   *D*

After reading the following material, the reader shall:

13.8   Construct the load line on a given set of output characteristics of a common emitter amplifier for a stated value of load resistance.

13.9   Estimate from the load line under no input signal conditions for given quiescent conditions the following:

(a)   voltage across transistor V_{CE}
(b)   voltage across load resistor
(c)   current through transistor I_C
(d)   current through load resistor
(e)   power dissipated in transistor
(f)   power dissipated in the load
(g)   power taken from the supply

These objectives are best fulfilled by considering the following measurements taken in the laboratory to find the output characteristics for an n–p–n transistor connected in common emitter configuration.

collector emitter voltage (V)	collector current (mA)				
	$I_B = 20\,\mu A$	$I_B = 40\,\mu A$	$I_B = 60\,\mu A$	$I_B = 80\,\mu A$	$I_B = 100\,\mu A$
1	0·6	2·4	4·2	6·0	7·8
3	1·0	2·8	4·6	6·4	8·2
5	1·4	3·2	5·0	6·8	8·6
7	1·8	3·6	5·4	7·2	9·0
9	2·2	4·0	5·8	7·6	9·4

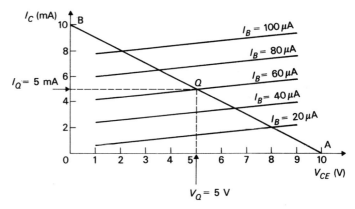

Figure 123 Operated under d.c. conditions.

These are plotted on the graph shown in Figure 123. The first step is to consider how the load line is plotted. For this purpose assume for the transistor used that $V_{CC} = 10\,V$ and the load resistor $R_L = 1000\,\Omega$.

Now $V_{CC} = I_C R_L + V_{CE}$

when $I_C = 0$ then $V_{CE} = V_{CC} = 10\,V$ (point A)

when $V_{CE} = 0$ then $I_C = \dfrac{V_{CC}}{R_L} = \dfrac{10}{1000} = 10\,mA$ (point B)

Points A and B are marked on the graph and joined to form the load line. The mid-point of this line occurs where $I_B = 60\,\mu A$ (approximately) and this is taken to be the operating or quiescent point Q. (In practice this point would be fixed by the biasing conditions.)

From the graph:

      Voltage across the transistor $V_Q = 5\,V$
∴   Voltage across the load R_L is $V_{CC} - V_Q = 10 - 5 = 5\,V$
      Current through transistor $I_Q = 5\,mA$
∴   Current through load R_L is $I_Q = 5\,mA$

Also total power dissipated in transistor is

$$V_Q \times I_Q = 5 \times 5 \times 10^{-3}$$
$$= 25\,mW$$

total power dissipated in load R_L is

$$(V_{CC} - V_Q) \times I_Q$$
$$= 5 \times 5 \times 10^{-3}$$
$$= 25 \, \text{mW}$$

total power taken from supply

$$V_{CC} \times I_Q = 10 \times 5 \times 10^{-3} = 50 \, \text{mW}$$

Note: Power from the supply is the power dissipated in the load plus the power dissipated in the transistor.

Self-assessment question

Figure 124 Characteristics for Question 13.

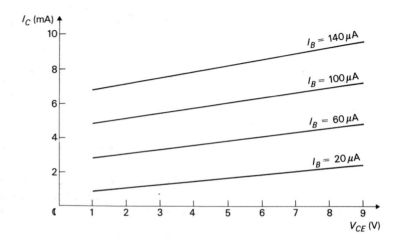

13   The output characteristics of a silicon transistor are shown in Figure 124. The transistor is connected in the circuit diagram shown in Figure 125.

Construct the load line on the output characteristics showing all the working involved. The bias current is set to 80 µA. Estimate from the graph the quiescent current and voltage and mark on the graph the operating point.

With no input signal applied, using the graph, find values for:

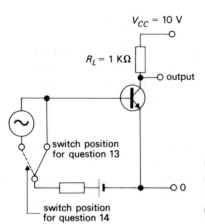

Figure 125 Circuit diagram for Question 13.

(a)   voltage across transistor V_{CE}
(b)   voltage across load resistor
(c)   current through load resistor
(d)   collector current I_C
(e)   power dissipated in transistor
(f)   power dissipated in load
(g)   power supplied from source

After reading the following material, the reader shall:

13.10 Estimate from the load line under sinusoidal input conditions only for given quiescent conditions the following:

(a) voltage across the transistor (collector–emitter)
(b) voltage across the load resistor
(c) current through the transistor (collector)
(d) current through the load resistor
(e) power dissipated in the transistor
(f) power dissipated in the load
(g) power taken from the supply
(h) voltage gain A_v
(i) current gain A_i
(j) power gain A_p in dB
(k) power dissipated at the collector junction of the transistor

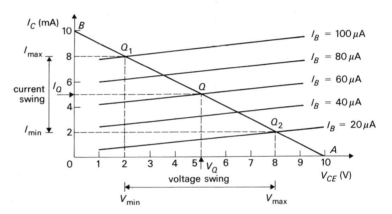

Figure 127 Operated under sinusoidal input conditions.

Consider the output characteristics shown in Figure 127. These are drawn with the load line already constructed. If a sinusoidal input waveform, of peak to peak value 80 µA is applied to the base of the transistor, the operating point Q moves to points Q_1 and Q_2. The peak value of the sinusoidal input waveform is 40 µA, causing the quiescent value of $I_B = 60$ µA to change from

point Q_1 given by $(60+40)$ µA $= 100$ µA
to point Q_2 given by $(60-40)$ µA $= 20$ µA

These points are marked on the graph shown in Figure 127 and give:

point Q_1 where $I_{max} = 8$ mA and $V_{min} = 2$ V
point Q_2 where $I_{min} = 2$ mA and $V_{max} = 8$ V

This gives the waveshapes shown in Figure 128.

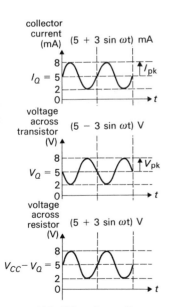

Figure 128 Waveshapes from graph.

Solution to self-assessment question (page 110)

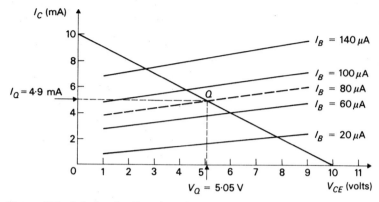

Figure 126   Solution for Question 13.

13   Now $V_{CC} = I_C R_L + V_{CE}$

$\therefore$ $10 = I_C 10^3 + V_{CE}$

When $I_C = 0$ then $V_{CE} = 10\,V$ to give point A

When $V_{CE} = 0$ then $I_C = \dfrac{10}{10^3} = 10\,mA$ to give point B

From these the load line is drawn.

Assuming the curves are equidistant from each other, $I_B = 80\,\mu A$ can be sketched in as a dotted line. This fixes the point Q.

$\therefore$ $V_Q = 5{\cdot}05\,V$ and $I_Q = 4{\cdot}9\,mA$

From the graph
- (a) $V_{CE} = 5{\cdot}05\,V$
- (b)   voltage across load resistor $= 4{\cdot}95\,V$
- (c)   current through load resistor $= 4{\cdot}9\,mA$
- (d) $I_C = 4{\cdot}9\,mA$
- (e)   Power dissipated in transistor $= 5{\cdot}05 \times 4{\cdot}9 = 24{\cdot}7\,mW$
- (f)   Power dissipated in load $= 4{\cdot}95 \times 4{\cdot}9 = 24{\cdot}3\,mW$
- (g)   Power supplied from source $= 10 \times 4{\cdot}9$ $= 49\,mW$

Now supply voltage = voltage across resistor + voltage across transistor.

From Figure 128 the supply voltage can be seen to be constant. In this case it is 10 V.

supply voltage $= V_{CC}$

Average power supplied $= V_{CC} I_Q = 10 \times 5 \times 10^{-3} = 50\,mW$

The average power in the load can be shown to be

$$(V_{CC} - V_Q)I_Q + \frac{V_{pk} I_{pk}}{2}$$

From Figure 128 the average power in the load is thus

$$(10-5) \times 5 \times 10^{-3} + \frac{3 \times 3}{2} \times 10^{-3} = 29 \cdot 5 \, \text{mW}$$

The first part is sometimes called the d.c. power and the second part the a.c. power. So,

d.c. power dissipated in load $= (V_{CC} - V_Q)I_Q$

a.c. power dissipated in load $= V_{pk}I_{pk}/2$

If the average power supplied by the source remains constant and the average power in the resistor increases by an amount $V_{pk}I_{pk}/2$ then the average power in the transistor must decrease by this amount when a sinusoidal signal is applied at the input to the amplifier.

This is true and the transistor runs cooler when an input signal is applied than when there is no input.

$\therefore$ Average power in transistor $=$ average power supplied

$-$ average power in load

$= (50 - 29 \cdot 5) \, \text{mW}$

$= 20 \cdot 5 \, \text{mW}$

A quick way to find the equivalent a.c. power developed in the circuit is to read from the graph and substitute in

$$\frac{(V_{max} - V_{min})(I_{max} - I_{min})}{8} = \frac{(8-2)(8-2) \times 10^{-3}}{8} = 4 \cdot 5 \, \text{mW}$$

If the input signal applied is sinusoidally varying, then the required output from the amplifier should also be a pure sinusoidal waveform that is increased in amplitude. Thus for comparison purposes to measure this increase the current gain is given by:

$$A_i = \frac{\text{change in collector current}}{\text{change in base current}}$$

or

$$A_i = \frac{I_{max} - I_{min}}{\text{change in base current}} = \frac{(8-2) \times 10^{-3}}{(100-20) \times 10^{-6}} = 75$$

To find the voltage gain it is necessary to know the value of the input resistance to the varying signal applied to the transistor. A typical value could be $1000 \, \Omega$.

$$A_v = \frac{V_{max} - V_{min}}{\text{change in base current} \times \text{resistance}}$$

$$\therefore A_v = \frac{8-2}{(100-20) \times 10^{-6} \times 10^3} = 75$$

The power gain is given by

$$A_p = A_v \times A_i$$
$$\therefore \quad A_p = 75 \times 75$$
$$\therefore \quad A_p = 5625, \text{ which is between } 10^3 \text{ and } 10^4$$

Thus putting A_p into a power form means it can be rewritten as $A_p = 10^{3 \cdot 75}$ which is a very useful way of dealing with these numbers. In practice the number 5625 is written as $10 \times 3 \cdot 75$ and then called a logarithmic power unit. The conversion process is shown in more detail as follows.

Power gain (expressed in logarithmic power units called dB or decibel)

$$= 10 \log_{10} A_p$$
$$= 10 \log_{10} 5625$$
$$= 37 \cdot 5 \, \text{dB}$$

Most of the heat dissipated in a transistor is dissipated at the collector junction, the remaining small amount at the emitter junction. If the base emitter voltage is assumed to be constant when an input signal is applied to a silicon transistor this is approximately $0 \cdot 6 \, \text{V}$.

Thus the average power dissipated at the emitter junction will be approximately $0 \cdot 6 I_Q$ watt. In the example this will be $0 \cdot 6 \times 5 \times 10^{-3} = 3 \, \text{mW}$.

$\therefore$ Average power dissipated at the transistor collector is given by $(25 - 4 \cdot 5 - 3) \, \text{mW}$ or $17 \cdot 5 \, \text{mW}$ when the input signal is applied.

Self-assessment question

14   A sinusoidal input of peak value $60 \, \mu\text{A}$ is applied to the input of the transistor in Question 13, which under these conditions has an a.c. input resistance of $900 \, \Omega$.

Sketch using estimated values taken from the graph the waveforms of

(i)   the transistor collector emitter voltage
(ii)   the voltage across the load resistor
(iii) the collector current

Estimate also from the graph

(a)   the power dissipated in the transistor
(b)   the power dissipated in the load
(c)   the power taken from the source
(d)   the power dissipated at the collector junction of the transistor
(e)   the voltage gain A_v
(f)   the current gain, A_i
(g)   the power gain, A_p, in dB

After reading the following material, the reader shall:

13.11   Describe thermal runaway of a transistor.

13.12   State the reason for the use of heat sinks.

When a transistor is switched on the power dissipated at the collector junction causes the junction temperature to rise. This causes an increase in leakage current. Hence the collector current rises above its quiescent value, which increases the power dissipated at the collector junction and causes a further increase in junction temperature. This process is accumulative, and could lead to the destruction of the device; then it is called thermal runaway. In small signal amplifiers careful design keeps it to a minimum. In large signal amplifiers it can be a problem.

When a collector current flows in a properly designed transistor circuit, at some time after switch on, the heat being lost by the transistor just equals the heat being generated within the transistor. Heat sinks are normally used to conduct this heat away from the transistor to the air surrounding the transistor.

A heat sink can be used:

(i)   to limit the temperature rise of a transistor by artificially increasing its size, hence allowing a higher value of collector current to be used

(ii)   to prevent the thermal runaway that would occur if no heat sink were used

(iii)   to ensure transistor insensitivity to ambient temperature changes, e.g. in constant temperature ovens

13.13   Sketch the circuit diagram and explain the action of a simple bias arrangement consisting of a resistor connected between V_{CC} and base.

13.14   Sketch the circuit diagram and explain the action of the following simple bias arrangements:

(*a*)   resistor connected between V_{CC} and base, and an emitter resistor

(*b*)   as in (*a*) with a decoupling capacitor

(*c*)   using a potential divider and an emitter resistor with a decoupling capacitor

13.15   Identify the difference between the d.c. and the dynamic or a.c. load line.

The bias arrangement used for the circuit shown in Figure 122 involves the use of a battery and a variable resistor R. Can the value of bias be set using other methods and if so what advantages or disadvantages do these circuits have? Consider the circuit diagram shown in Figure 131.

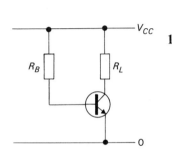

Figure 131   Simple biasing circuit.

Solution to self-assessment question (page 114)

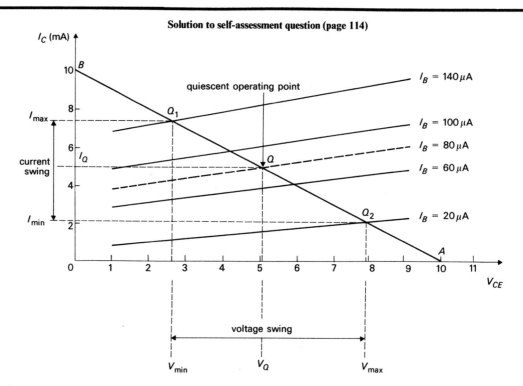

Figure 129   Solution for Question 14.

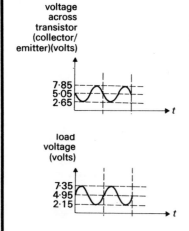

Figure 130 `Sketches of waveform for
Question 14.

14   The load line is constructed as shown in Question 13 under input signal conditions:

Using the input quoted gives a maximum base current of $80 + 60 = 240\,\mu A$ and hence point Q_1. The minimum base current is given by $80 - 60 = 20\,\mu A$ and hence point Q_2 as shown in Figure 129.

Points Q_1 and Q_2 give the values of

$$V_{max} = 7\cdot85\,V \text{ and } V_{min} = 2\cdot65\,V$$
$$I_{min} = 2\cdot1\,mA \text{ and } I_{max} = 7\cdot3\,mA$$

Hence the waveforms are as shown in Figure 130.

From these it can be seen that the waveforms are not symmetrical about the effective d.c. value. The original assumption that the curves were equidistant apart was not true; however, this error can be assumed to be negligible.

From Question 13 under no input signal conditions,

Power dissipated in transistor = 24·7 mW
Power dissipated in load   = 24·3 mW
Power supplied from source   = 49 mW

Now the effective a.c. power dissipated is

$$\frac{(V_{max} - V_{min})(I_{max} - I_{min})}{8} = \frac{(7\cdot85 - 2\cdot65)(7\cdot3 - 2\cdot1)}{8}$$
$$= 3\cdot38\,mW$$

(a)   Power dissipated in transistor $= 24 \cdot 7 - 3 \cdot 38 = 21 \cdot 32 \, \text{mW}$

(b)   Power dissipated in load $= 24 \cdot 3 + 3 \cdot 38 = 27 \cdot 68 \, \text{mW}$

(c)   Power taken from source $= 27 \cdot 68 + 21 \cdot 32 = 49 \, \text{mW}$

(d)   Power dissipated at collector $= 24 \cdot 7 - 3 \cdot 38 - I_Q(0 \cdot 6)$

          junction of transistor $= 21 \cdot 32 - 0 \cdot 6 \times 4 \cdot 9$

                                 $= 18 \cdot 38 \, \text{mW}$

(e)   Voltage gain,

$$A_V = \frac{V_{max} - V_{min}}{(\text{change in base current}) \times 900}$$

$$\therefore \; A_V = \frac{7 \cdot 85 - 2 \cdot 65}{(140 - 20) \times 10^{-6} \times 900}$$

$$= 48$$

(f)   Current gain,

$$A_I = \frac{I_{max} - I_{min}}{\text{change in base current}} \quad \therefore \; A_I = \frac{(7 \cdot 3 - 2 \cdot 1) \times 10^{-3}}{140 - 20) \times 10^{-6}}$$

$$= 43 \cdot 3$$

(g)   Power gain,

$$A_p = 48 \times 43 \cdot 3$$

$$\therefore \; A_p = 2080$$

$$\therefore \; \text{Gain in dB} = 10 \log_{10} 2080 = 33 \cdot 2 \, \text{dB}$$

By careful choice of the value of the resistor R_B the bias current can be set to give the required quiescent conditions. Consider how the value of R_B is found if a silicon transistor is used.

Now $V_{CC} = 6\,V$ and $h_{FE} = 100$

It is required to find R_B for a collector current of $1\cdot5\,mA$.

From the circuit diagram shown in Figure 131

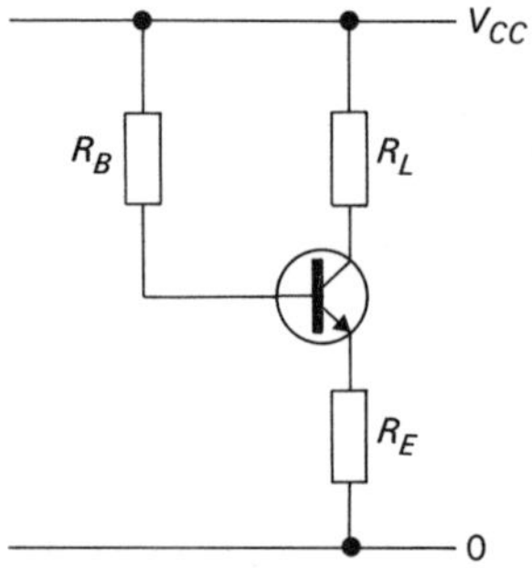

Voltage across

$$R_B = 6 - 0\cdot6 = 5\cdot4\,V$$

Also

$$\frac{I_C}{I_B} = h_{FE}$$

$$\therefore\ I_B = \frac{1\cdot5 \times 10^{-3}}{100} = 15\,\mu A$$

$$\therefore\ R_B = \frac{5\cdot4}{15 \times 10^{-6}} = 360\,k\Omega$$

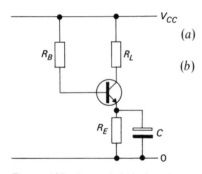

Figure 132   Biasing circuit.

However the circuit has the following disadvantages

(a)   The collector current is directly proportionate to h_{FE}. Therefore the circuit will not withstand changes in h_{FE}.

(b)   A small change in the leakage current produces a large change in I_C.

Thus a change in temperature is accumulative, which can lead to collector current variations and hence some circuit arrangement is required to stabilize the circuit against changes in the d.c. collector current.

Consider the circuit shown in Figure 132 where the collector current is stabilized under d.c. conditions.

An increase of collector current is accompanied by an almost equal increase of emitter current. This results in an increase of voltage developed across the emitter resistor reducing the forward bias of the emitter base junction. The base current is reduced causing a decrease in the collector current compensating for the original increase. This reduces however the gain measured for equivalent a.c. signals.

To avoid this a capacitor is connected across R_E which acts as a short circuit at the frequency of the a.c. signal as shown in Figure 133. Thus at d.c. levels the working point is stabilized and yet gain is stabilized for the a.c. signals being amplified.

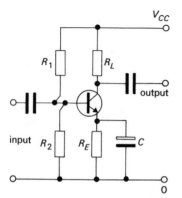

Figure 133   Improved biasing circuit.

A more comprehensive circuit is shown in Figure 134.

The potential divider chain R_1 and R_2 maintains a constant voltage between base and zero potential. The best d.c. stability is obtained for low values of R_1 and R_2, but this has the following disadvantages.

Figure 134   Circuit of small signal amplifier.

(*a*) A large d.c. current is drawn.

(*b*) R_1 and R_2 appear in parallel across the input for a.c. signals, and hence reduce the input resistance of the stage to a.c. signals.

A rough guide is to make the current in R_1 and R_2 to be approximately ten times the highest expected base current. As shown in Figure 134, the emitter resistor and capacitor function in the same way as previously described.

The value of R_E chosen is also a compromise. It must be sufficiently large to provide adequate temperature compensation but small enough to keep power loss to a minimum. In practice $R_E \approx 1/I_C$ is a convenient value.

In this case the equation relating the collector current I_C is

$$V_{CC} = I_C(R_L + R_E) + V_{CE}$$

This gives a load line of slope $-1/(R_L + R_E)$ and hence the quiescent point Q can be fixed as shown in Figure 135.

The capacitor C causes the varying part of the emitter current to bypass the resistor R_E. Thus the a.c. resistance in series with the transistor is R_L assuming that the source resistance is zero. This gives an equivalent circuit for the varying part of the current as shown in Figure 136.

Thus the point Q is fixed using the load line of slope $-1/(R_L + R_E)$, called the d.c. load line.

The stage gain is computed from the load line of slope $-1/R_L$ called the dynamic or a.c. load line as shown in Figure 135.

The dynamic load line is dependent on the type of circuit used. The original circuit used in Figure 114 has a d.c. load line that is the same as the a.c. load line but as can be seen this is *not* the case for Figure 135. Further considerations of dynamic load lines are reserved for other units.

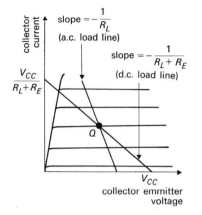

Figure 135   D.c. and a.c. load line.

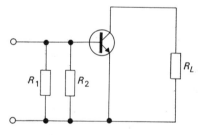

shunt input circuit reducing
input resistance

Figure 136   Equivalent circuit.

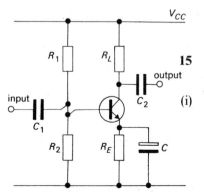

Figure 137   Circuit diagram for Question 15.

Self-assessment questions

15   Tick the correct alternative for each part. The parts relate to the circuit diagram shown in Figure 137.

(i)   If R_L went open circuit:

A The amplifier gain would fall to zero.

B The transistor would overheat, leading to thermal runaway.

C The output signal would suffer from amplitude distortion.

D The transistor bias would increase.

(ii) If *C* went open circuit:
  A The output signal would fall to zero.
  B The collector current would be zero.
  C The transistor would overheat leading to thermal runaway.
  D The amplifier gain would be reduced.

(iii) If the voltage measured across R_2 and the voltage measured across R_E were both zero the fault could be:
  A R_1 has gone open circuit.
  B *C* has gone short circuit.
  C Transistor has an internal short between base and emitter.
  D R_L has gone open circuit.

16 Sketch the circuit diagram of the simple biasing arrangement consisting of a resistor connected between V_{CC} and base for a common emitter amplifier.

(i)   State the main disadvantage of this circuit.
ii)   Explain how the circuit can be improved using another resistor connected between emitter and ground.
(iii) State the main disadvantage of this improvement.
(iv) Explain how the circuit could now be improved using a capacitor.
(v)   Sketch the final improved circuit for biasing.

17 As can be seen from Question 16(v) the simple circuit has become more complex. The sinusoidally varying part of the output signal can have a different effective circuit resistance to that of the steady part of the signal. Complete the following sentences.

(i)   The d.c. load line is used to establish the _____ point.
(ii) The dynamic or a.c. load line is used to estimate the _____ of the amplifier.

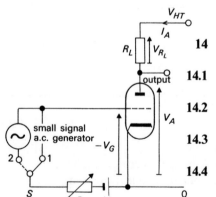

Figure 140 Thermionic tride valve amplifier.

After reading the following material, the reader shall:

14 Know the circuit and operation of a small signal thermionic triode amplifier.
14.1 Sketch the circuit diagram of a single stage voltage amplifier having a load resistor and with all direct voltages and currents indicated.
14.2 Explain that bias is required to obtain a selected quiescent operating point on the output characteristics.
14.3 Explain the effect of a small sinusoidal voltage input on the quiescent conditions.
14.4 State that voltage inversion occurs between input and output signals.

Consider the circuit shown in Figure 140. When the switch is set to position 1, adding the voltages around the supply part of the circuit gives

Equation 2 $$V_{H.T.} = I_A R_L + V_A$$

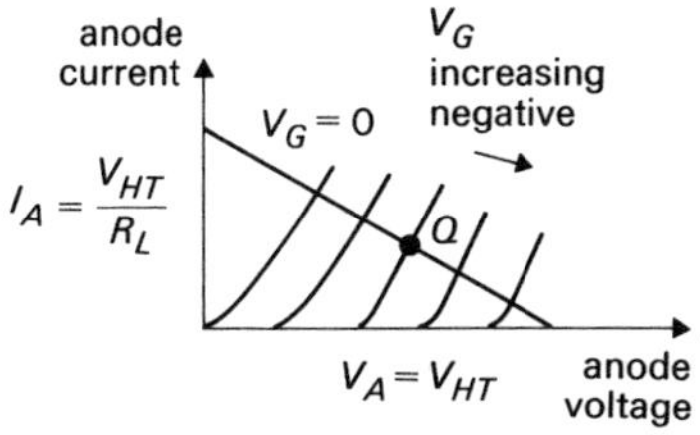

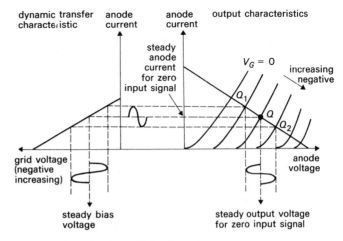

anode current

$$I_A = \frac{V_{HT}}{R_L}$$

V_G increasing negative

$V_G = 0$

Q

$V_A = V_{HT}$ anode voltage

Figure 141 Load line plotted on output characteristics.

This gives the equation of the d.c. load line which can be plotted in the usual way on the output characteristics as shown in Figure 141.

Varying the value of R fixes the quiescent operating point Q as shown on the graph. In practice other methods are used to apply the bias without the need to use a separate bias voltage V.

If the switch in Figure 140 is then switched to position 2 a varying sinusoidal signal is applied to the grid of the valve. This causes the point Q to move up and down the load line as shown in Figure 142.

dynamic transfer characteristic · anode current · anode current · output characteristics

steady anode current for zero input signal

$V_G = 0$ · increasing negative

Q_1 · Q · Q_2

grid voltage (negative increasing) · anode voltage

steady bias voltage · steady output voltage for zero input signal

Figure 142 Effect of a sinusoidal voltage input.

From Equation 2 it can be seen that if the anode current increases, then the voltage across the load resistor increases. However, as the supply voltage is constant the voltage across the valve decreases. A series of waveforms can be drawn for the voltage and current waveforms as was done for the transistor in Figure 120. It is found that when the input voltage to the grid *increases* the anode to cathode voltage of the valve *decreases*.

Thus the output signal voltage is the *inverse* of the input signal voltage.

After reading the following material, the reader shall:

15 Construct and use a d.c. load line on a thermionic triode characteristic.

15.1 Construct the load line for a stated value of load resistance on a given set of output characteristics of a triode amplifier.

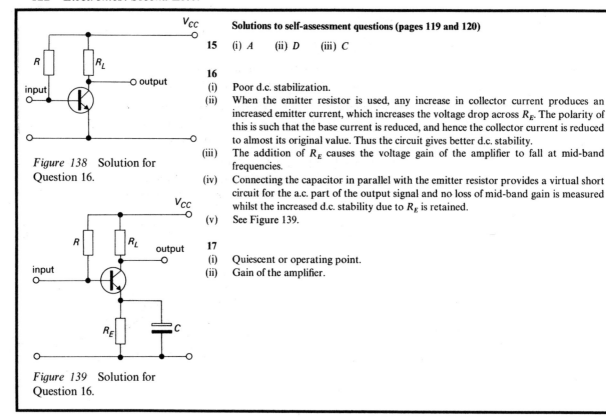

V_{CC}

R   *R$_L$*

input

output

Figure 138   Solution for Question 16.

V_{CC}

R   *R$_L$*   output

input

R$_E$ *C*

Figure 139   Solution for Question 16.

Solutions to self-assessment questions (pages 119 and 120)

15   (i) *A*   (ii) *D*   (iii) *C*

16
(i)   Poor d.c. stabilization.
(ii)   When the emitter resistor is used, any increase in collector current produces an increased emitter current, which increases the voltage drop across R_E. The polarity of this is such that the base current is reduced, and hence the collector current is reduced to almost its original value. Thus the circuit gives better d.c. stability.
(iii)   The addition of R_E causes the voltage gain of the amplifier to fall at mid-band frequencies.
(iv)   Connecting the capacitor in parallel with the emitter resistor provides a virtual short circuit for the a.c. part of the output signal and no loss of mid-band gain is measured whilst the increased d.c. stability due to R_E is retained.
(v)   See Figure 139.

17
(i)   Quiescent or operating point.
(ii)   Gain of the amplifier.

15.2   Select a suitable value of bias voltage to obtain a desired quiescent point on the output characteristics.

15.3   Estimate from the load line under no input signal conditions for given quiescent conditions the following:
(*a*)   power taken from the supply
(*b*)   power dissipated in the load
(*c*)   power dissipated in the valve

15.4   Estimate from the load line under sinusoidal input conditions only for given quiescent conditions the following:
(*a*)   r.m.s. value of the equivalent a.c. part of the voltage output signal
(*b*)   voltage gain A_v of the stage
(*c*)   equivalent a.c. power developed by the stage

15.5   Explain the operation of a typical small signal thermionic triode amplifier.

These objectives are best considered by using measurements taken in the laboratory and applying them. Consider the following table of results which were taken for a circuit as shown in Figure 140.

anode voltage (V)	anode current (mA)				
	$V_G = 0\,\text{V}$	$V_G = -2\,\text{V}$	$V_G = -4\,\text{V}$	$V_G = -6\,\text{V}$	$V_G = -8\,\text{V}$
50	3	0	0	0	0
100	9	3	0	0	0
150	15	9	3	0	0
200	21	15	9	3	0
250		21	15	9	3
300			21	15	9

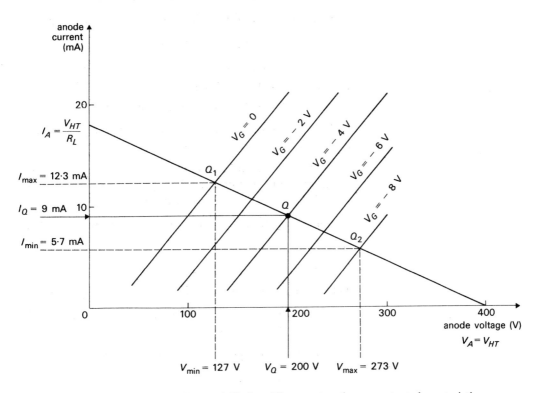

Figure 143 Load line construction on output characteristics.

These are plotted on the graph shown in Figure 143. The first step is to consider how the load line is plotted. For this purpose assume that for the valve used, that the supply voltage is 400 V and the value of the load resistor is measured to be 22·2 kΩ.

Now $V_{H.T.} = I_A R_L + V_A$

when $I_A = 0$ then $V_A = V_{H.T.} = 400\,\text{V}$

when $V_A = 0$ then $I_A = \dfrac{V_{H.T.}}{R_L} = \dfrac{400}{22\cdot2 \times 10^3} = 18\,\text{mA}$

This fixes the two end points of the load line on the graph. What would be a suitable value of bias? Providing that Q_1 does not move beyond the line given by $V_G = 0$ and the characteristics are evenly spaced apart, Q can be fixed approximately half way along the load line. In this case a suitable value of V_G would by $V_G = -4\,\text{V}$, and the maximum value of applied varying signal (so that no distortion occurs) would be the peak value of the sinusoidal wave of 4 V.

In practice the curves are not linear or equally spaced, and the best point Q must be chosen with care, at the same time considering the maximum value of signal to be applied at the input.

From the graph shown in Figure 143 with *no signal* applied:

$$V_Q = 200\,\text{V} \quad \text{and} \quad I_Q = 9\,\text{mA}$$

Power taken from supply $= V_{H\,T}.I_Q = 400 \times 9 \times 10^{-3} = 3{\cdot}6\,\text{W}$
Power dissipated in load $= I_Q^2 R_L = (9 \times 10^{-3})^2 \times 22 \times 10^3 = 1{\cdot}8\,\text{W}$
Power dissipated in valve $= V_Q I_Q = 200 \times 9 \times 10^{-3} = 1{\cdot}8\,\text{W}$

Note: Power taken from supply is the power dissipated in the load plus the power dissipated in the valve.

If now a sinusoidal input signal of peak value 4 V is applied to the grid of the valve the amplifier is being operated under signal conditions. The 4 V has been chosen to illustrate the example.

The operating point Q moves to the extremes on the load line of

Q_1 given by $V_G = -4 + 4 = 0\,\text{V}$ and
Q_2 given by $V_G = -4 - 4 = -8\,\text{V}$

Thus V_A changes from 273 to 127 V as the input signal varies.

∴ The equivalent a.c. part of the output voltage varies peak to peak by $273 - 127 = 146\,\text{V}$

∴ Converting to a peak value gives $\frac{146}{2} = 73\,\text{V}$

∴ Converting to an r.m.s. value gives $73 \times 0{\cdot}707 = 51{\cdot}6\,\text{V}$

The voltage gain

$$A_v = \frac{\text{change in anode voltage}}{\text{change in grid voltage}}$$

$$\therefore \ A_v = \frac{273 - 127}{0 - (-8)} = 18{\cdot}25$$

The equivalent a.c. power developed is given by:

$$\frac{(V_{\text{max}} - V_{\text{min}})(I_{\text{max}} - I_{\text{min}})}{8}$$

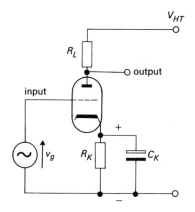

Figure 144 Small signal thermionic triode valve amplifier.

From the graph this is given by

$$\frac{(273-127)(12{\cdot}3-5{\cdot}7)\times 10^{-3}}{8} = 120{\cdot}5\,\text{mW}$$

Other required quantities can be found using similar principles to those described for the transistor.

For example, the average power dissipated by the load resistor *increases* by 120·5 mW when the signal is applied, and that the average power dissipated by the valve *decreases* by 120·5 mW when the signal is applied.

In the previous consideration as to how bias is provided, a battery and resistor were used. Another method that avoids this requirement is to use the parallel combination of R_K and C_K as shown in the circuit diagram of Figure 144. The operation of the circuit using R_L is as previously described except in the way in which bias is applied. If C_K was not included, then a potential difference would be measured across R_K when a signal is applied at the input. The polarity of this potential is as shown on the diagram. The function of C_K is to provide a low reactace path for the varying part of the anode current, and hence ensure a constant value of voltage gain at mid-band frequencies for the amplifier.

If the value of $R_K = 1\,\text{k}\Omega$ and the d.c. component of the anode current was 5 mA the grid bias would be -5 V.

If $C_K = 50\,\mu\text{F}$ then its reactance $= 1/2\pi fC_K$

$$\therefore\ X = \frac{1}{2\pi f \times 50 \times 10^{-6}} = \frac{10^4}{\pi}\times\frac{1}{f}\,\text{ohm}$$

For a frequency of 100 Hz this reactance is given by 32 Ω.

Also it can be seen from this formula that further increase of frequency causes a reduction in the reactance of the capacitor.

Hence R_K can be assumed to be bypassed as far as the varying component of the anode current is concerned.

The use of this combination means that

(i) A *d.c.* load line must be used to fix the *operating point*.
(ii) A *dynamic (or a.c.)* load line must be used to calculate the *gain*.

Self-assessment questions

18 Sketch the circuit diagram of a single stage thermionic triode voltage amplifier having a load resistor with all direct voltages and currents indicated. The diagram should include:

(*a*)   input and output terminals
(*b*)   load resistor
(*c*)   bias resistor
(*d*)   decoupling capacitor
(*e*)   supply with polarity indicated.

19   Complete the following sentences either by writing in the correct word or by deleting the incorrect word where applicable.

(*a*)   The purpose of using bias in a single stage thermionic-triode voltage amplifier is to obtain on the output characteristics a selected

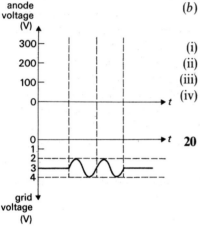

(*b*)   Applying a small increasing sinusoidal signal to a single stage thermionic-triode voltage amplifier causes the following:

(i)   the anode current increases/decreases.
(ii)   the voltage across the load resistor increases/decreases.
(iii) the output voltage signal increases/decreases.
(iv)   the output signal voltage is the inverse of the _____ signal voltage.

Figure 145 Waveshape for Question 20.

20   In a thermionic-triode resistive loaded amplifier, the voltage gain is measured to be 100 when the sinusoidal signal shown in Figure 145 is applied at the input. When the signal is removed the voltage measured at the output is a steady 200 V. This assumes the valve is correctly biased. Sketch the waveshape of the voltage measured at the output, marking on the maximum and minimum values of voltage relating their position on the time axis shown in Figure 145 to that of the input signal.

21

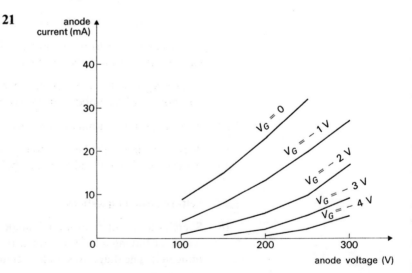

Figure 146 Characteristics for Question 21.

The output characteristics of a thermionic-triode are shown in Figure 146. The triode is then used as part of a resistive-loaded voltage amplifier, where the load resistance is measured to be 7·5 kΩ, and the supply source to be used is 300 V.

Construct the load line on the output characteristics showing all the working involved.

If the input peak value of the sinusoidal signal is never allowed to rise above 1 V select the value of grid bias to give minimum distortion and maximum amplification.

22 For the amplifier of question 21, the value of grid bias is taken to be $V_G = -1$ V. For a sinusoidal input signal of peak to peak value 2 V estimate from the graph constructed for Question 21:

(*a*) the r.m.s. value of the varying part of the output voltage (the effective a.c. component of the output voltage)
(*b*) the power dissipated in the load
(*c*) the voltage gain

Post test – small signal amplifiers

23 A silicon transistor is biased by connecting a resistor R_B between $V_{CC} = 9$ V and the base of the transistor.

It is required to produce a collector current of 2·5 mA and the H_{FE} of the t ansistor can be assumed to be 100.

Calculate the value of R_B required to produce this value of collector current.

24 The following results were obtained from a transistor connected in a common emitter configuration.

V_{CE} (V)	I_C (mA)		
	$I_B = 0\,\mu$	$I_B = 40\,\mu$A	$I_B = 80\,\mu$A
1	0·4	2·6	5·1
5	0·45	2·8	5·6
10	0·55	3·0	6·1

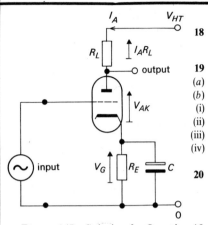

Figure 147   Solution for Question 18.

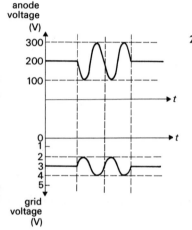

Figure 148   Solution for Question 20.

Solutions to self-assessment questions (pages 124–126)

18   See Figure 147.

19
(a)   quiescent point *or* operating point.
(b)
(i)   increases
(ii)   increases
(iii)   decreases
(iv)   input

20   See Figure 148.

Voltage gain $= 100$

$\therefore \ 100 = \dfrac{V_{max} - V_{min}}{4 - 2}$

$\therefore \ V_{max} - V_{min} = 200$ centred at 200 V.

The signal is the *inverse* of that at the grid.

21   To plot the load line then

$$V_{H.T.} = I_A R_L + V_A$$
$$\therefore \ 300 = I_A \times 7{\cdot}5 \times 10^3 + V_A$$

when

$I_A = 0,$ then $V_A = 300\,\text{V}$ giving point A

$V_A = 0,$ then $I_A = \dfrac{300}{7{\cdot}5 \times 10^{-3}}$

$\therefore \ I_A = 40\,\text{mA}$ giving point B

It can be seen that the characteristics increase in equal increments of grid voltage. However the spacing is not equal between the lines.

If $V_G = -1\,\text{V}$ is taken as the bias value a change of bias will produce a larger change in anode voltage than if $V_G = -3\,\text{V}$ was taken because the spacing between lines changes.

A transistor with a collector load of $1{\cdot}25\,\text{k}\Omega$ is connected in the common emitter configuration with a steady base current set at $40\,\mu\text{A}$ and a supply voltage of 8 V. From the above data plot the I_C/V_{CE} characteristics and draw on the load line. From the graph estimate:

(i)   The quiescent current and voltage
(ii)   The power taken from the supply

A sinusoidal signal of r.m.s. value $28{\cdot}28\,\mu\text{A}$ is applied at the base.

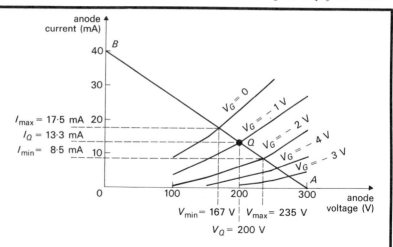

If $V_G = -2\,\mathrm{V}$ was taken then the movement of Q about the 2 V line would be unequal for equal changes of grid bias causing distortion of the output signal.

A compromise would appear to be $V_G = -1\,\mathrm{V}$ where distortion is kept to a minimum and amplification is maximized.

22   From the graph $V_Q = 200\,\mathrm{V}$ and $I_Q = 13\cdot3\,\mathrm{mA}$

$$\text{effective a.c. power} = \frac{(V_{\max}-V_{\min})(I_{\max}-I_{\min})}{8}$$

$$= \frac{(235-167)(17\cdot5-8\cdot5)}{8} = 76\cdot5\,\mathrm{mW}$$

(b) ^{2} Power dissipated in load $= [(300-200)(13\cdot3)+76\cdot5]\,\mathrm{mW}$
$$= 209\cdot5\,\mathrm{mW}$$

(a)   The r.m.s. value of $V_a = \dfrac{V_{\max}-V_{\min}}{2\sqrt2} = \dfrac{(235-167)}{2\sqrt2} = 24\,\mathrm{V}$

(c)   Voltage gain $A_v = \dfrac{V_{\max}-V_{\min}}{\text{change in input voltage}} = \dfrac{235-167}{2} = 34.$

From the graph estimate:
(a)   Maximum and minimum voltage at the collector
(b)   Maximum and minimum current through the load
(c)   Voltage gain if the a.c. input resistance to the transistor is $1\,\mathrm{k\Omega}$
(d)   current gain
(e)   power gain in dB
(f)   Reduction in power dissipated in transistor when the input sinusoidal signal is applied.

25   The following figures were obtained from measurements made to determine the static anode characteristics of a triode valve.

anode voltage, V_A(V)		10 20 30 40 50 60 70 80 90 100 110 120
anode current I_A (mA)	$V_G = 0$ V	0·7 2·4 4·8 7·3 9·8
	$V_G = -1$ V	0·6 2·4 4·8 7·4 9·9
	$V_G = -2$ V	0·2 1·0 2·8 5·2 7·6
	$V_G = -3$ V	0·6 1·8 3·4 5·8

The valve is used with an anode load of 10 kΩ and a supply voltage of 120 V. Plot the static anode characteristics and the load line. From the graph estimate:

(*a*)   the quiescent voltage and current for $V_G = -1·5$ V
(*b*)   the power taken from the supply using the bias of (*a*)
(*c*)   the voltage amplification for small changes in the grid voltage for $V_G = -1·5$ V.

State any assumptions made.

Topic area Waveform generators

After reading the following material, the reader shall:

17 Know the principle of simple oscillator operation.

17.1 State the purpose of an oscillator.

17.2 State that an oscillator must consist of two parts, namely the frequency determining unit and the loss replacement unit.

The purpose of an oscillator is to produce an output signal that:

(i) changes magnitudes between stated levels

(ii) produces a recurring waveshape at the required frequency

The frequency and the waveshape chosen for oscillation depend mainly on the purpose for which the output signal is required.

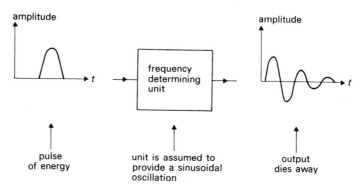

Figure 150 Frequency determining unit.

Take the diagram shown in Figure 150. If a pulse of energy is applied to the frequency determining unit, the unit oscillates, but due to practical considerations this natural oscillation always dies down to zero. There are always *associated losses* in this part. If sufficient energy is applied to the oscillatory unit to overcome these losses, and if it is applied at an appropriate time in the cycle, oscillations continue indefinitely to provide a *maintained oscillatory unit*, shown in Figure 151.

This addition of energy is provided by the *loss replacement unit*.

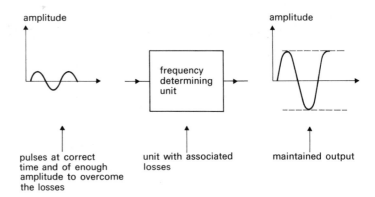

amplitude

amplitude

frequency
determining
unit

pulses at correct
time and of enough
amplitude to overcome
the losses

unit with associated
losses

maintained output

Figure 151 Maintained oscillatory unit.

Solutions to post test (pages 127–30)

23 Assume that for a silicon transistor $V_{BE} = 0\cdot6$ V

^{2} Voltage across R_B

is $9 - 0\cdot6 = 8\cdot4$ V

Also

$$I_B = \frac{I_C}{h_{FE}} = \frac{2\cdot5 \times 10^{-3}}{100}$$

$$= 25\cdot0\,\mu\text{A}$$

$$\therefore\ R_B = \frac{8\cdot4}{25 \times 10^{-6}} = 336\,\text{k}\Omega$$

24

(i) $2\cdot75\,\text{mA}$ $4\cdot55\,\text{V}$

(ii) $12\cdot65\,\text{mW}$

(a) $7\cdot4\,\text{V}$ $1\cdot55\,\text{V}$

(b) $5\cdot15\,\text{mA}$ $0\cdot5\,\text{mA}$

(c) $73\cdot6$

(d) $58\cdot1$

(e) $36\,\text{dB}$

(f) $3\cdot42\,\text{mW}$

25

(a) From a sketch of $V_G = -1\cdot5\,\text{V}$

$V_Q = 73\cdot5\,\text{V}$ and $I_Q = 4\cdot7\,\text{mA}$

(b) $0\cdot35\,\text{W}$

(c) Assume an input signal (sinusoidal) of peak value $1\cdot5$ V to swing Q to known values on the curve. Hence $A_v = 19\cdot7$.

After reading the following material, the reader shall:

17.3 Identify the type of oscillator by the method of loss replacement.

17.4 Sketch output waveforms of oscillators in common use: sinusoidal, rectangular, sawtooth.

17.5 Identify the terms period, frequency, magnitude, peak value, mark, space, mark to space ratio, where applicable to the waveforms in 17.4.

17.6 State the common uses of the waveforms set out in 17.4.

Switching

The energy loss in the frequency determining unit is replaced from a d.c. source where an *electronic switch* ensures that the energy is applied at the correct time in the oscillatory cycle. It normally consists of two independent circuits such that the output of each circuit controls the input of the other as shown in Figure 152. This type is known as a *relaxation oscillator*.

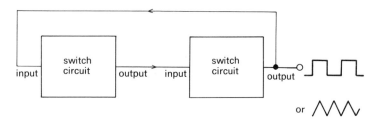

Figure 152 Relaxation oscillator.

Feedback

The output from the frequency determining unit is fed back to its own input in such a sense that the feedback signal aids the change of signal at the input. The energy loss is supplied by an amplifier connected as in Figure 153. In this case no input signal is required as the frequency determining unit provides its own signal via the feedback network. This type is known as a *feedback* oscillator.

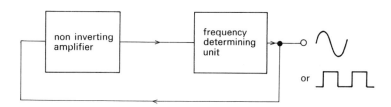

Figure 153 Feedback oscillator.

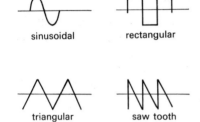

sinusoidal   rectangular

triangular   saw tooth

Figure 154   Types of oscillator waveforms.

Negative resistance

The oscillations which would normally die away in a non-maintained circuit due to resistance could continue indefinitely if the resistance was eliminated. The loss replacement section can be regarded as negative resistance; when added to the resistance of the frequency determining unit it just cancels it out, thus providing the conditions for maintained oscillation. This type is known as a *negative resistance* oscillator.

Typical waveforms obtained from oscillators are shown in Figure 154.

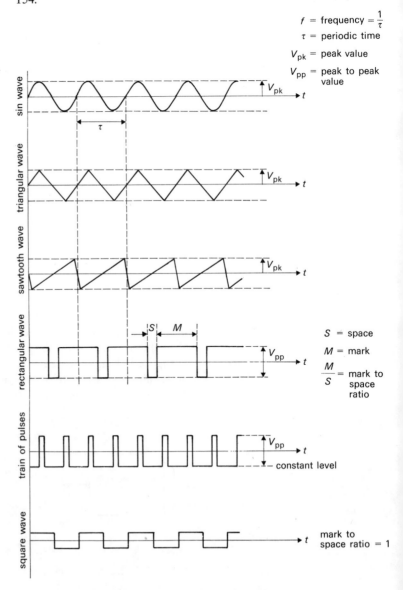

f = frequency = $\frac{1}{\tau}$

τ = periodic time

V_{pk} = peak value

V_{pp} = peak to peak value

S = space

M = mark

$\dfrac{M}{S}$ = mark to space ratio

Figure 155   Sketches of waveforms.

The sawtooth waveform is a particular type of triangular waveform.

The square wave is a particular type of rectangular waveform.

Consider the waveform shown in Figure 155. The *amplitude*, the instantaneous magnitude measured with respect to a reference level, varies continuously. However, it can be seen that the pattern repeats itself in a time called the *periodic time*, the time taken for one complete cycle. The symbol for this is the Greek letter tau or τ and the units are seconds. The *maximum amplitude* reached in each cycle is called the *peak value*. In many cases the positive peak value is numerically equal to the negative peak value, but this is not always so. The number of cycles occurring continuously in one second is called the *frequency*. In particular the rectangular waveform can be made irregular as shown in Figure 155 such that the time the pulse operates at one level is called the mark, and the time the pulse operates at the other level is called the space. The *mark to space ratio* shows how much faster the switching process of one of the oscillator switches operates compared to the other. When the mark to space ratio is equal to one the rectangular waveform is known as a square wave.

Typical uses of waveforms shown in Figure 155 are as follows:

Relaxation oscillator

These are based around the monostable, bistable and astable devices. The use of 'mono' and 'bi' describes the number of available stable states in which the circuit stays until an external trigger pulse is applied. Thus the monostable can be in one stable state only, and the bistable can be in either of two stable states only. The astable circuit has no stable states and switches continuously between states. (Further treatment of these circuits is considered in topic area Oscillators in *Electronics 3*.)

When used to produce square or rectangular wave shapes then:

(*a*)   The monostable is used for shaping or generating pulses.
(*b*)   The bistable is used as a two-state device for logic applications.
(*c*)   The astable is used for pulse train generators in logic application, especially in computers. It is also used for many other electronic purposes e.g. for tone generation (with the requisite filters) in musical circuits.

The astable device (or multivibrator) can be used with a converting circuit to produce triangular or sawtooth waveforms. These can be used as time base generators for oscilloscope or television circuits. This is done by using a direct output from a relaxation oscillator (of the thyristor type) or by using a multivibrator oscillator and a

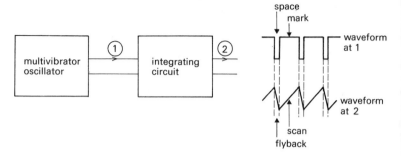

Figure 156 Time base generator.

converting circuit as shown in Figure 156. The circuit is known as an integrating circuit (further details are given in *Electronics 3*.) It can be seen that this is the waveform required to provide the scanning voltage to the *X* plates of the cathode ray tube.

Feedback oscillator

These are normally arranged to provide a sinewave output. They are used in transmitters to provide the carrier frequency for the modulator, and in receivers to provide the carrier frequency for the demodulator. They are also used in test instruments where the generation of sinewaves is required.

Negative resistance oscillator

These are also used where sinewave outputs are required. A typical example is the use of a crystal oscillator and a tunnel diode to counteract the energy loss within the crystal.

Table 10 gives an indication of the names of oscillators relating to the frequency range over which the oscillator is used.

oscillator	range
audio frequency (a.f.)	few Hz to 20 kHz
video frequency (v.f.)	audibility to low MHz
radio frequency (r.f.)	above audibility
high frequency (h.f.)	very high and ultra high frequency bands

Table 10 Oscillators and frequencies.

Self-assessment questions

Complete the following sentences related to oscillators either by writing in the correct word or deleting the incorrect word where applicable.

1 The purposes of an oscillator is to produce an output signal that:
(*a*) changes magnitude/frequency between stated levels
(*b*) produces a random/recurring waveshape at the required frequency.

2 The type of oscillator that applies feedback from the output of an amplifier to aid the amplifier input signal via a frequency determining unit is called a _____ oscillator.

3 The type of oscillator that consists of two interdependent circuits such that the output of each circuit controls the input of the other is called a _____ oscillator.

4 The type of oscillator that has a loss replacement unit which has a negative resistance is known as a _____ oscillator.

5 Sketch and name the output waveforms that would be obtained from the oscillators of Questions 2, 3, and 4.

6 Match the numbers on the waveshape shown in Figure 157 with the relevant term where possible.

A periodic time
B peak value
C mark
D space
E frequency
F mark to space ratio

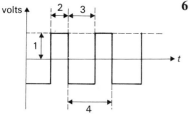

Figure 157 Waveshape for Question 6.

7 State two common uses of the:

(*a*) feedback oscillator
(*b*) relaxation oscillator

After reading the following material, the reader shall:

18 Know the principles of simple sinusoidal oscillators.
18.1 Identify the difference between positive and negative feedback applied to amplifiers.
18.2 State that a sinewave oscillator is an amplifier with positive feedback sufficient to maintain its own input.
18.3 State that a sinewave oscillator requires both a frequency determining circuit and a method of self stabilization.

Feedback is said to occur when some of the output of a system is fed back to the input.

When a small proportion of the output signal of an amplifier is fed back to the input, as shown in Figure 159, the feedback signal tends to *aid* the input signal. This results in the two signals adding and a larger output signal. The gain of the amplifier has been *increased*. The amount of increase that can be obtained in this way is very limited because only a small amount of feedback can be applied without causing the amplifier to become unstable and oscillate.

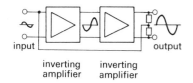

Figure 159 Feedback signal aiding the input signal.

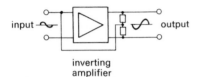

Figure 160   Feedback signal opposing the input signal.

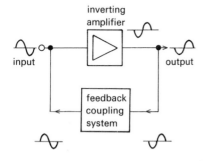

Figure 161   Positive feedback using a feedback coupling system.

This type of feedback is known as *positive feedback* (or regeneration). It is rarely used to increase an amplifier gain, but is a basic requirement of sinewave oscillators.

When a small proportion of the output signal of an amplifier is fed back to the input, as shown in Figure 160, the feedback signal tends to *oppose* the input signal. This results in the two signals subtracting and a reduced output signal. The gain of the amplifier has been *reduced*.

This type of feedback is known as *negative feedback* (or degeneration) and it would seem that there is little use in applying negative feedback. However it can be seen from topic area Feedback in *Electronics 3* that other benefits can be obtained from negative feedback.

The positive feedback system shown in Figure 159 uses two inverting amplifiers to provide the right kind of signal at the input. If only one inverting amplifier is used, the feedback coupling system is made to invert the output waveform so that it aids the input signal, ensuring positive feedback, as shown in Figure 161. This can be done by using either a transformer or a phase shift network.

To obtain the required inversion for positive feedback to aid the input signal, either two inverting amplifiers are required, as shown in Figure 159, or one inverting amplifier with a feedback coupling system, as shown in Figure 161. Therefore in practice either of these systems has *associated inductance* and *capacitance*. What effect does this have on

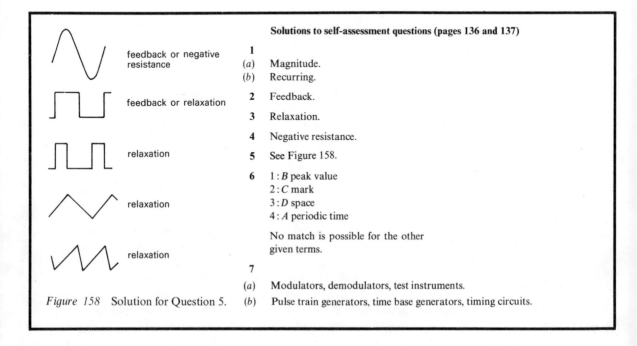

Figure 158   Solution for Question 5.

Solutions to self-assessment questions (pages 136 and 137)

1
(*a*)   Magnitude.
(*b*)   Recurring.

2   Feedback.

3   Relaxation.

4   Negative resistance.

5   See Figure 158.

6   1 : *B* peak value
2 : *C* mark
3 : *D* space
4 : *A* periodic time

No match is possible for the other given terms.

7
(*a*)   Modulators, demodulators, test instruments.
(*b*)   Pulse train generators, time base generators, timing circuits.

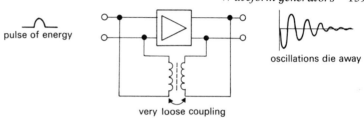

Figure 162 Positive feedback with very loose coupling.

the system? Consider the diagram shown in Figure 162, where the coupling system is of the transformer type.

If a pulse of energy is applied at the input when very loose coupling is used, the amplifier oscillates at the frequency determined by the associated circuit components, but the feedback is so weak that oscillations are not maintained.

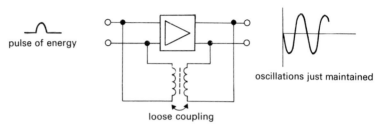

Figure 163 Positive feedback with loose coupling.

If a pulse of energy is applied at the input as shown in Figure 163, oscillations occur. If the coupling of the transformer is reduced, it is found that at one point the oscillations die away. Thus *to maintain oscillations* the coupling of the transformer must be greater than this value. If the oscillations are not started by this pulse of energy, tighter coupling must be used to be certain that oscillations start.

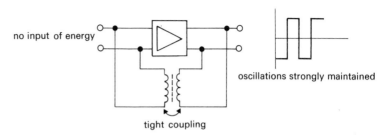

Figure 164 Positive feedback with tight coupling.

If *very tight coupling* is used, as shown in Figure 164, the output from the amplifier, which is driven hard on or off, will approximate to a square wave, and no input energy is required to start oscillations.

However with this system there is no circuit or unit to determine the frequency at which the system oscillates.

The oscillation is determined by the associated circuit reactances which are impossible to pre-determine.

If the associated reactances are pre-determined (and not left to chance circuit reactances) the frequency of the output sinewave of an oscillator can be more carefully controlled. Thus both the feedback and the negative resistance oscillator require *frequency determining networks* to be of practical use. Typical methods of providing this circuit for the feedback oscillator are shown in Figure 165. Note also that the order of the coupling system and the frequency determining network of the LC system can be reversed.

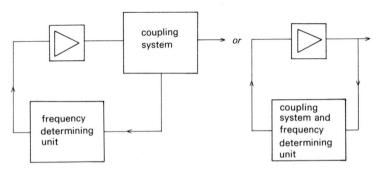

basic system for LC oscillators   basic system for RC oscillators

Figure 165 Oscillator using a frequency determining circuit.

In the negative resistance oscillator a crystal can be used as a frequency determining circuit. When connected to a device that exhibits negative-resistance (a tunnel diode, say), the natural oscillations, which can be accurately pre-determined, are maintained. Each crystal has one characteristic frequency pre-determined by its physical characteristics, and cannot be used at any other frequency.

It can be seen from Figure 166 that *external influences* affect the *internal performance* of the oscillator. The frequency of oscillation depends mainly on the frequency determining network, but other factors can make it vary. To improve the frequency stability the following can be used.

Frequency determining unit
Use components that have small temperature coefficients.
Use a crystal in a temperature controlled environment.
Use a crystal to control a tuned circuit.

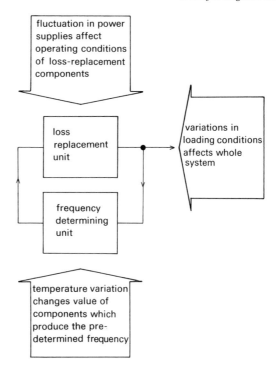

Figure 166 Causes of frequency instability in oscillators.

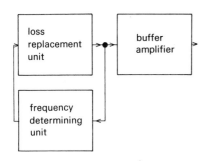

Figure 166 Causes of frequency instability in oscillators.

Loss replacement unit
Use stabilized power supplies.

Variation in loading
Use a buffer amplifier, which isolates the oscillator from any changes in load, and increases the power level.

Thus to provide a sinewave output of stable frequency it is necessary to use the system shown in Figure 167, which uses all the methods previously indicated to improve frequency stability.

Self-assessment questions

Complete the following sentences related to oscillators either by writing in the correct word or deleting the incorrect word where applicable.

8 When a small proportion of the output signal of an amplifier is coupled back to the input signal such that it aids the input signal, negative/positive feedback is applied.

9 When a small proportion of the output signal of an amplifier is coupled back to the input signal such that it opposes the input signal, negative/positive feedback is applied.

10 A sinewave oscillator is an amplifier with positive feedback sufficient to _____ its own output.

11 State the two basic requirements of a sinewave oscillator that ensure its output signal has a known stable frequency.

After reading the following material, the reader shall:

18.4 Describe the oscillatory process of a coil and an inductor connected in parallel.

18.5 State the requirements of a tuned LC circuit used in a feedback oscillator.

18.6 State that the approximate frequency of oscillation of most LC oscillators is

$$f_0 = \frac{1}{2\pi\sqrt{LC}}.$$

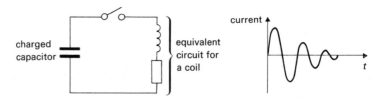

Figure 168   Charged capacitor connected to a coil.

Take the circuit shown in Figure 168. If the capacitor is initially charged to a potential and connected across a coil, a current flows which produces a magnetic field around the inductor. When the capacitor is discharged, the magnetic field collapses and induces an e.m.f. in the inductor. This e.m.f. causes a current to flow in the opposite direction. This sets up an electric field between the plates of the capacitor. *Energy is continually converted from a magnetic field to an electric field and vice-versa.* The oscillations die away as energy is dissipated in the associated resistance of the circuit. If the circuit had no resistance then no energy would be lost, but this never occurs under normal operating conditions. The energy lost by resistance has to be replaced if oscillations are maintained. The frequency at which oscillations occur is called the *natural frequency* of the circuit. It depends on the resistance, inductance and capacitance of the circuit. A graph of current against time for the circuit is shown in Figure 168.

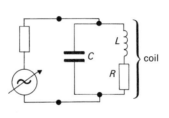
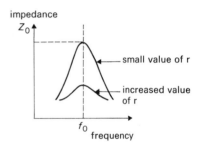

f_0 is the resonant frequency
Z_0 is the dynamic impedance

Figure 169   Response of the tuned LC circuit.

The associated resistance of the coil is made small in practice to achieve minimum loss of energy. This has another effect, which can be seen from Figure 169. To maintain oscillations in this circuit the applied positive feedback should be the same frequency as the natural frequency of the circuit. If a variable frequency is applied to the circuit and a graph of the impedance plotted against frequency (the voltage of the generator is kept constant) the shape of the graph is as shown in Figure 169. If the resistance of the coil is increased the peak reduces and the graph widens about the resonant frequency.

Assuming the currents supplied by different frequencies are of equal amplitude, the signal voltages developed across the tuned circuit ($V = IZ$) are directly proportional to the impedance of the circuit at each frequency. Hence the voltage developed by a current I at the resonant frequency i.e. IZ_0 will be at a maximum. The voltages developed by the currents at other frequencies is much smaller. The more narrow the graph the greater is the falling off of voltage magnitudes close to the resonant frequency. Another way of saying this is that *these voltages are rejected*, and the circuit is made more selective by decreasing the value of coil resistance.

Thus for high selectivity and low loss the resistance of the coil should be low compared with the reactance of the coil. A measure of this ratio is the Q factor of the coil which is given by

$$Q = \frac{\text{reactance of coil}}{\text{resistance of coil}}.$$

A requirement of r.f. tuned circuits is coils with high Q factors.

It becomes more difficult to construct a high Q a.f. coil as the frequency decreases because of the component values required. At these frequencies, unless carefully designed, a.f. feedback oscillators can produce a distorted output due to poor selectivity.

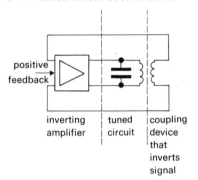

Figure 170 Tuned circuit feedback oscillator.

As shown in Figure 170, the tuned circuit provides a voltage of known frequency across a known load resistance. The purest sinewave is obtained with 'loose' coupling.

If the amount of positive feedback is progressively increased the output rises but becomes more distorted. Thus although the LC circuit can be used for a.f. oscillators it is normally used for r.f. oscillators. Pure a.f. waveforms are generally obtained using the RC phase shift networks, as shown previously in Figure 165.

The LC tuned circuit is required to have good selectivity. If the coil resistance is small compared to its reactance it can be shown that the frequency of oscillation or resonant frequency is given by:

$$f_0 = \frac{1}{2\pi\sqrt{LC}}$$

where

L = inductance of coil in henrys (H)

C = capacitance of capacitor in farads (F)

A wide variety of feedback oscillators have been developed using the LC tuned circuit. The position in which the tuned circuit appears in the feedback oscillator, and the way it is used, gives rise to typical oscillators a few of which are:

> tuned grid oscillator
> tuned collector oscillator
> Meissner oscillator
> Hartly oscillator
> Colpitts oscillator

All of these oscillators can be shown to have a frequency of oscillation given by:

$$f_0 = \frac{1}{2\pi\sqrt{LC}}$$

where L and C are obtained from an individual analysis of the circuit used.

Solutions to self-assessment questions (pages 141 and 142)

8   Positive.

9   Negative.

10 Maintain.

11 Frequency determining circuit. Method of self stabilization.

After reading the following material, the reader shall:

18.7 Sketch the circuit diagram of a tuned collector oscillator and a thermionic triode oscillator.

18.8 Describe in simple terms the operation of a tuned oscillator and a tuned grid oscillator.

18.9 Describe methods of applying bias in a tuned-circuit transistor oscillator.

18.10 State the two conditions required for maintained oscillations in a feedback oscillator.

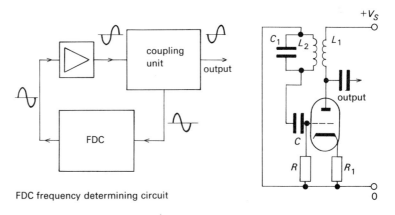

FDC frequency determining circuit

Figure 171 Circuit diagram of tuned grid oscillator.

Consider the circuit shown in Figure 171. The amplifier is acting as the loss replacement unit providing energy via the coupling system to the tuned circuit of L_1 and C_1. The coupling systems depends on the mutual inductance between L_1 and L_2. What is the purpose of the other components C, R and R_1? Consider the part of the circuit shown in Figure 172.

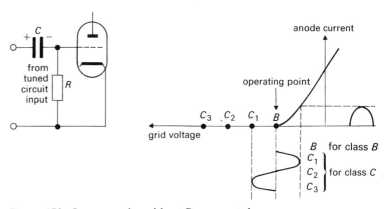

Figure 172 Input to valve without R_1 connected.

At the instant the oscillator is switched on the input to grid of the valve is zero. The signal from the tuned circuit then establishes the bias level. On positive half cycles the grid to cathode part of the valve behaves as a diode, and grid current flows, charging the capacitor with the polarity shown. On negative half cycles the capacitor discharges via R, but not by the same amount. The forward resistance of the diode on charge is low, but the reverse resistance of the diode on discharge is high. The bias voltage (on the capacitor) builds up on the first few cycles of input signal until equilibrium conditions are reached, such that the charge gained during the positive cycle equals the charge lost on the negative cycle. The amount that the operating point moves to the left depends on the signal strength. The values of R and C are carefully chosen to ensure that enough bias is applied automatically to keep the operating point where it is shown on the Figure 172 or to some point to the left of this point. This means the anode current consists of half or less than half pulses of energy to feed the tuned circuit. The valve is said to be operated in class B conditions if the operating point is as shown, or if the operating point is to the left of this the valve is said to be operating in class C conditions. R_1 is included to prevent excessive current through the valve at the instant of switch on, ensuring cathode bias and hence protecting the device.

Distortion of the output does not occur (as it would in a class C audio amplifier), because waveform purity at the output of the oscillator is maintained by the natural cycling of the tuned circuit.

Consider the circuit diagram, shown in Figure 173, of the tuned collector oscillator. The principle of operation is basically the same as for the tuned grid circuit except the coupling unit and the frequency determining circuit are interchanged. The transistor is being used as a common emitter amplifier, where the parallel combination of L_1 and

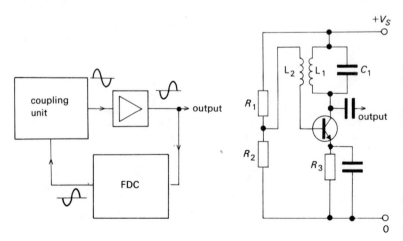

Figure 173   Tuned collector oscillator.

C_1 form the frequency determining circuit, and mutual coupling from L_1 and L_2 provides a path for feedback from the output to the input. Coupling must be tight enough to ensure oscillations do not die away, but not so tight that distortion occurs.

The other difference is in the method of biasing the transistor. The potential divider R_1 and R_2 keeps the voltage between the base and zero rail fairly constant provided the base current is a small proportion of the current through R_1 and R_2. The emitter resistor R_3 prevents changes in collector current as the temperature varies.

If the collector current rises due to temperature change, the voltage across R_3 increases. This means the potential difference between the base and emitter is reduced, and the collector current is reduced. This form of feedback is negative feedback which when used in this way is called d.c. feedback. (For further details see topic area Feedback in *Electronics 3*.) To prevent the fall in gain which would occur at the oscillatory frequency (by inserting this resistor) a decoupling capacitor C_3, which has almost zero reactance at the oscillatory frequency, is connected across R_3. This has no effect on R_3 at d.c. levels. When the amplifier is operated in this way it is said to be biased under class A conditions, shown in Figure 174.

It should be noted that the values of R_1 and R_2 fix the working point of the transistor, and hence the biasing mode of the transistor.

If the transistor is used to reproduce the input signal to the base exactly at the collector (class A conditions), the bias is made

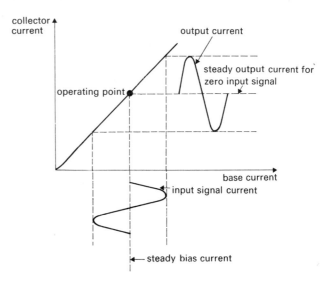

Figure 174 Tuned collector oscillator operated under class A conditions.

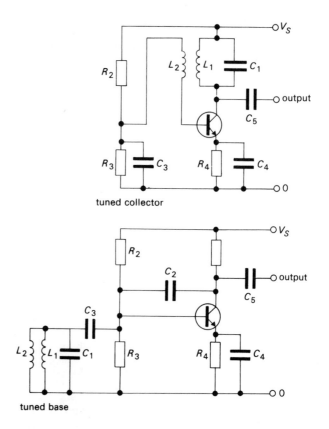

 tuned collector

tuned base

Figure 175 Examples of tuned circuit transistor oscillators.

dependent on the circuit components. This is shown in Figure 175, where the resistors R_2 and R_3 are used to fix the operating point of the tuned collector transistor; R_4 applies d.c. negative feedback to prevent drift of the operating conditions; and C_3 and C_4 provide a direct path to the zero rail at the oscillatory frequency of operation. In the tuned base transistor, C_3 is required to prevent the bias for the transistor being primarily determined by the low d.c. resistance of winding L_1. Feedback from the output to the input is provided by C_2 and L_2.

If the type of bias applied to a tuned circuit transistor oscillator causes pulses of output current to appear at the collector of the transistor, then the bias has to be arranged such that the operating point operates the transistor under class B or class C conditions. If auto bias is used it is dependent on signal strength, but if standard methods using circuit components are used this is not the case.

Consider the diagram shown in Figure 176.

The feedback applied must be positive feedback to aid the input signal.

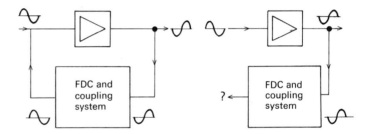

Figure 176 Feedback oscillator.

If the connection between the frequency determining circuit and the coupling system is broken, what are the requirements of the signal to be fedback with the input signal? They are:

(i) The signal fed back should be of such a phase (related to the input signal) *that it aids the input signal.*

(ii) The aiding signal fed back should have enough magnitude to ensure that *maintained oscillations* take place.

The first point implies that the output signal from the loss replacement unit must be corrected in phase such that it aids the input to the frequency determining circuit.

The second point implies that if an amplifier is used, the gain of the amplifier multiplied by the combined attenuation of the frequency determining circuit and the coupling system, must be greater than unity. Another way of saying this is that the open loop gain is greater than unity.

Note: An amplifier increases the magnitude of an electrical signal. An attenuator decreases the magnitude of an electrical signal.

Self-assessment questions

12 Tick the correct statements. It may be assumed that the statements occur in a logical order, in a recognized time sequence.

If a capacitor is initially charged to a known potential and then connected across a coil:

(*a*) a current flows in a particular direction
(*b*) a magnetic field is set up around the inductor
or
(*c*) an electric field is set up around the inductor
(*d*) the inductor's associated field collapses
(*e*) a current flows in the opposite (as opposed to the previous) direction
(*f*) a magnetic field is set up across the capacitor

or

(g)   an electric field is set up across the capacitor
(h)   energy is being continually interchanged between the capacitance and the inductance and vice-versa
(i)   oscillations occur indefinitely
or
(j)   oscillations die away.

13   Give the reason for the answer that has been ticked from either (i) or (j).

Complete the following sentences by deleting the incorrect word or writing the formula where applicable.

14   For high selectivity in an LC tuned circuit the resistance/reactance of the coil should be low compared with the resistance/reactance of the coil.

15   The approximate frequency of oscillation of most LC oscillators is given by $f_0 = $ _____.

16   Sketch the circuit diagram of a tuned collector transistor oscillator and a tuned grid oscillator.

17   In simple terms describe how the circuits of Question 16 operate under the following headings:
(a)   frequency determining circuit
(b)   loss replacement unit
(c)   temperature stabilization
(d)   components to fix the operating point of the active device.

Post test – waveform generators
Tick the correct answer.

18 The purpose of a buffer amplifier connected to the output of an oscillator is to provide:

(*a*) the energy to maintain oscillations
(*b*) frequency stability as the temperature varies
(*c*) frequency stability as the loading condition varies
(*d*) inversion of the input signal and hence positive feedback to the input of the amplifier.

19 The type of oscillator generally used to produce a train of rectangular pulses is called a:

(*a*) feedback oscillator
(*b*) relaxation oscillator
(*c*) negative resistance oscillator
(*d*) sinewave oscillator

20 The type of oscillator used to provide the time base generator voltage for a cathode ray tube is called a:

(*a*) feedback oscillator
(*b*) relaxation oscillator
(*c*) negative resistance oscillator
(*d*) sinewave oscillator

21 Which of the following units are required to build a feedback oscillator that produces a sinewave output signal where frequency stability is not important?

(*a*) inverting amplifier, tuned circuit, buffer amplifier
(*b*) inverting amplifier, crystal, buffer amplifier
(*c*) non-inverting amplifier, tuned circuit
(*d*) inverting amplifier, tuned circuit

22 Sketch the waveforms stated, marking where applicable the period, peak value, mark and space:

(*a*) sinewave
(*b*) sawtooth wave
(*c*) rectangular wave
(*d*) square wave

23 State three common causes of frequency instability in LC feedback oscillators.

24 Give one method of improving frequency stability for each of the causes stated in Question 23.

Solutions to self-assessment questions (pages 149 and 150)

12   *a, b, d, e, g, h* and *j* are correct.

13   Oscillations die away because of the associated resistance of the coil and connecting leads. The answer assumes that the circuit is at room temperature, and not at very low temperatures.

14   Resistance, reactance.

15
$$f_0 = \frac{1}{2\pi\sqrt{LC}}$$

16

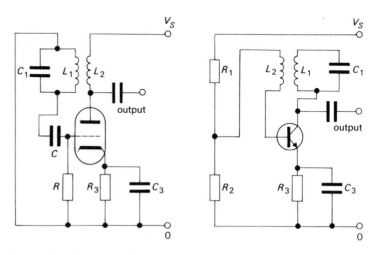

Figure 177   Solution to Question 16.

17

(a)   The frequency determining circuit $L_1 C_1$ is a high Q, tuned circuit, where a capacitor and a coil are connected in parallel.

(b)   The loss replacement unit is provided by the triode or the transistor being used as an amplifier.

(c)   A limited amount of temperature stabilization is probided by resistor R_3 which causes negative d.c. feedback to be applied. The capacitor C_3 decouples this resistor at the oscillatory frequency, so that at this frequency there is no loss of amplifier gain.

(d)   In the triode circuit C and R are used to provide automatic bias. This operates the triode under class C conditions, and thus pulses of anode current are provided to replace the energy loss of the tuned circuit.

In the transistor circuit R_1 and R_2 are chosen to provide a steady base current, and hence fix the operating class of the transistor. If the collector waveform is a pure reproduction of the input signal to the base, the transistor is operated under class A conditions.

Complete the following sentences by adding the correct word or words.

25   The purpose of a tuned circuit is to provide across a known load resistance, a voltage of known _____.

26   When the resistance of a coil used in a tuned circuit is made small compared with the reactance of the coil, the tuned circuit has high

_____.

27   Although the LC tuned circuit can be used for a.f. oscillators it is normally used for _____ oscillators.

28   Pure a.f. waveforms are generally obtained using an RC _____ _____ network.

29   For maintained oscillations in a feedback oscillator:

(*a*)   the signal fed back should be of such a phase relative to the input signal that it _____ the input signal

(*b*)   to ensure that maintained oscillations take place the signal fed back should have enough _____.

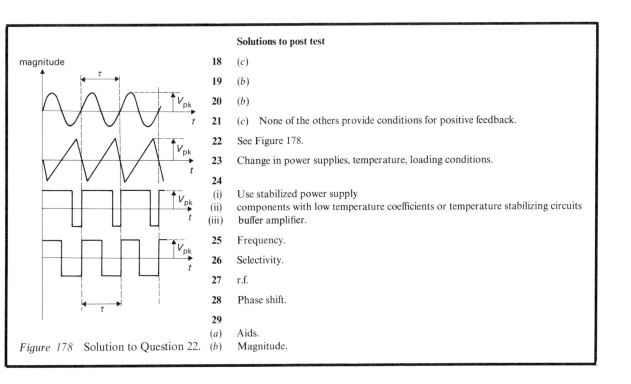

Figure 178   Solution to Question 22.

Solutions to post test

18   (*c*)

19   (*b*)

20   (*b*)

21   (*c*)   None of the others provide conditions for positive feedback.

22   See Figure 178.

23   Change in power supplies, temperature, loading conditions.

24
(i)   Use stabilized power supply
(ii)   components with low temperature coefficients or temperature stabilizing circuits
(iii)   buffer amplifier.

25   Frequency.

26   Selectivity.

27   r.f.

28   Phase shift.

29
(*a*)   Aids.
(*b*)   Magnitude.

Topic area Logic elements and circuits

After reading the following material, the reader shall:

19 Know that information can be communicated by two-state signals.
19.1 Identify an electrical communication system
(*a*) a transmitter and encoder
(*b*) a transmission medium
(*c*) a receiver and decoder
19.2 Give simple examples of two-state devices.
19.3 Give simple examples of information being communicated by two-state devices.
19.4 Recognize the difference between positive and negative logic.

Passing information from one point (the transmitting point) to another point (the receiving point) is known as *communication*. It requires the system to use a common language or *code*. Before the discovery of electricity communication was effective over comparatively short distances only.

One function of the *transmitter* is to *encode* the information to be passed in a suitable way for the *transmission medium*, the substance through which the code passes from transmitting point to receiving point.

The *receiver* must then *decode* the signal back into its original form. Early communication systems in which the information was encoded included beacons, smoke signals, flags or lanterns. They are not suitable for long distance transmission due to large *attenuation* in the transmission medium, which reduces the information signal magnitude.

The effectiveness of communication was improved by changing the transmission medium to copper wire. A simple example is shown in Figure 179.

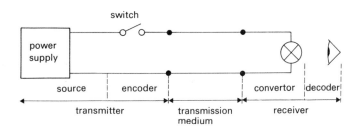

Figure 179   Simple communication system.

When the key (the encoder) is closed, current flows along the wire (the transmission medium) causing the lamp to light. The eye acts as the decoder providing the code is known.

A simple ON/OFF code would pass little information but a complex code would pass much more information. An example of this is the code developed by Samuel Morse, in which the coded signals take the form of combinations of short and long bursts of energy with critical spacings.

The lamp shown in Figure 180*a* is a *two-state device*. It is either on or off. (Instead of a lamp a buzzer could be used.)

Similarly a pen could be used which contacted the paper only when the current was flowing, as shown in Figure 180*b*. The pen could be made to deflect from its zero current point to a point of maximum current, as shown in Figure 180*c*. These examples depend on the fact that a switch can only be on or off.

Another method of changing the state of a device is to reverse the direction of current. This could cause a small ferrite core to change its direction of magnetization, as shown in Figure 180*d*. It could also change the direction of magnetization on a magnetic surface (e.g. magnetic tape), as shown in Figure 180*e*.

Figure 180   Examples of two-state devices.

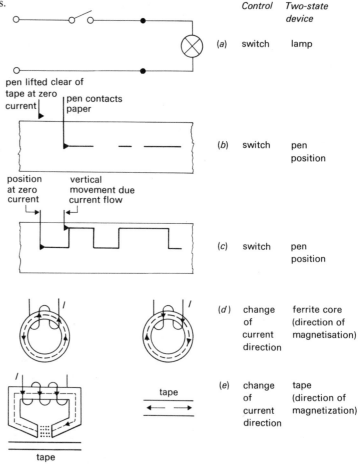

		Control	Two-state device
	(a)	switch	lamp
	(b)	switch	pen position
	(c)	switch	pen position
	(d)	change of current direction	ferrite core (direction of magnetisation)
	(e)	change of current direction	tape (direction of magnetization)

These examples depend on the *change of level of the signal* that is connected to the device that stores the information. When the energy source is switched off the information stored may be lost, e.g. the lamp which is controlled by a switching circuit.

Alternatively when the energy source is switched off the information stored may be retained, e.g. the magnetic surface which is controlled by a change in level of current.

In modern electronic systems a change in level of the signal controls the switching device which causes the storing device to change state.

The change in level is caused by a *train of electrical pulses* representing the information to be communicated.

The switching device is normally a high speed device, e.g. a transistor that switches on or off according to the electrical pulse level applied to the input.

Figure 181 Simple communication system.

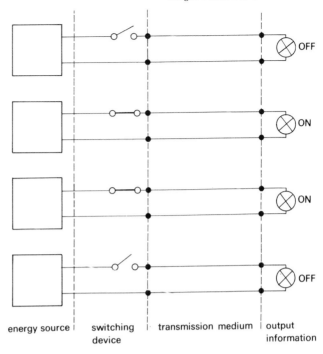

Figure 182 Electronic communication system.

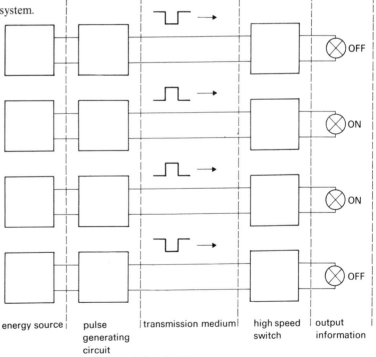

A positive pulse causes the switch to be ON.
A negative pulse causes the switch to be OFF.
The switch only changes over when a pulse is applied.

The storing device represents the information contained in the train of electrical pulses.

A simple system that transmits information, e.g. a code that represents a number by four different levels, is shown in Figure 181.

When the levels are transmitted instantaneously in this form, the information is said to be transmitted in *parallel*. The simple system could be converted to an electronic system as shown in Figure 182.

Note also that the only possible states of the output are on or off. Information represented in this form is called binary information. One disadvantage of the system shown in Figure 182 is that it needs four sets of circuits, which is expensive for long distance transmission, so it is usual to transmit binary information over one set of lines.

The same information is transmitted over one set of lines in *serial form* in Figure 183. However the information takes longer to be transmitted.

Figure 183 Serial transmission of binary information.

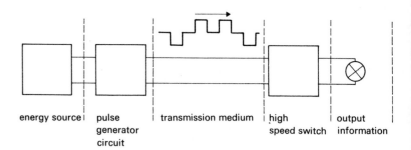

energy source | pulse generator circuit | transmission medium | high speed switch | output information

Parallel transmission of information is important in computer technology where high speed operation is more important than the extra cost of apparatus.

For the same information to be transmitted, in Figure 182 the lamps are, respectively, OFF, ON, ON, and OFF, all at once. In Figure 183 the single lamp is first OFF, it then goes ON, it remains ON, and then it goes OFF.

Figure 184 Closer packing of information.

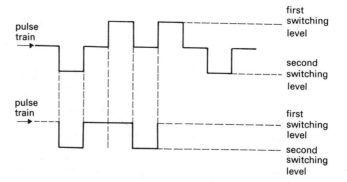

The pulses shown increase or decrease about a constant level. This is wasteful in terms of packing information into a fixed time. the pulses can be packed closer together, as shown in Figure 184, providing that the two levels operate the high speed switching device.

The differences in level of binary information are called '1' or '0'. This allows the levels to represent binary arithmetic. This means a row of '1's or '0's could represent words or numbers, depending on the information to be transmitted. These combinations of '0's or '1's can be represented in two alternative forms, as shown in Figure 185.

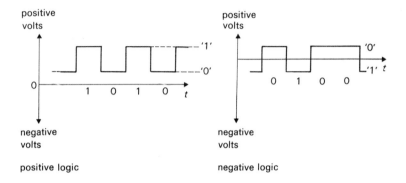

Figure 185 Negative and positive logic.

If the more *positive potential* represents the '1' the system is said to use *positive logic*.

If the more *negative potential* represents the '1' the system is said to use *negative logic*.

Self-assessment questions

Complete the following sentences.

1 Before the discovery of electricity, communication could take place over short distances only because the transmission medium had a high _____ factor.

2 The use of copper wires as the transmission medium enables information to be transmitted over _____ distances.

3 Draw the circuit diagram for a simple communication system that contains an energy source, switch, transmission medium and an output.

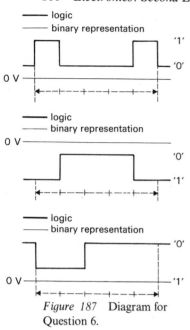

Figure 187   Diagram for
Question 6.

4   Give two examples of a two state device that

(*a*)   loses stored information when the controlling device is de-energized.
(*b*)   does not lose stored information when the controlling device is de-energized.

5   The pulses shown in Figure 186 are fed into a controlling device that operates a lamp.
Level '1' causes the lamp to light.
Level '0' causes the lamp to go out.
State how the lamp condition varies for the time interval *AB* shown in Figure 186.

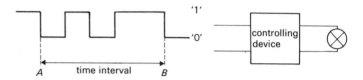

Figure 186   Diagram for Question 5.

6   In Figure 187 state in each case the form of logic which is in use, and in each case write the binary representation of the information.

After reading the following material, the reader shall:

20   Understand the function of 'AND', 'OR' and 'NOT' gates.
20.1   Explain how different combinations of three switches connected in series control the ON/OFF condition of a lamp.
20.2   Explain how different combinations of three switches connected in parallel control the ON/OFF condition of a lamp.
20.3   Explain how trains of pulses are controlled in electronic systems using logic gates.

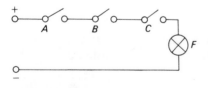

Figure 189   Series connection of switches.

Consider three switches *A*, *B* and *C* connected in series with a lamp *F* to a supply source as shown in Figure 189. The lamp *F* is only ON if all the switches *A* and *B* and *C* are ON.

If an ON condition is represented by a '1' (known as logic 1) and an OFF condition represented by a '0' (known as logic 0) Table 11 can be constructed to denote the lamp condition for all possible conditions or states of switches *A*, *B* and *C*.

A	B	C	F
0	0	0	0
0	0	1	0
0	1	0	0
0	1	1	0
1	0	0	0
1	0	1	0
1	1	0	0
1	1	1	1

Table 11 Possible switch positions (series).

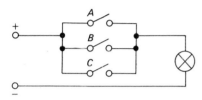

Figure 190 Parallel connection of switches.

A	B	C	F
0	0	0	0
0	0	1	1
0	1	0	1
0	1	1	1
1	0	0	1
1	0	1	1
1	1	0	1
1	1	1	1

Table 12 Possible switch positions (parallel).

Consider what now happens when the three switches are connected in parallel to a lamp *F* via a supply source, as shown in Figure 190.

The lamp will only be ON if either switch *A* or switch *B* or switch *C* is ON.

Table 12 shows the lamp condition for all possible conditions of switches *A*, *B* and *C*.

The switch allows the flow of current to reach the lamp. Opening a gate allows people to pass through. In the same way *electronic gates* are required that allow pulses to pass through. A particular combination of input signals may open a gate and allow an output signal to be measured. So different types of gates exist that respond to *different combinations* of input signals to obtain an output signal to be measured.

These gates are called as *logic gates*.

At the present time of writing it is intended by British Standards to bring the symbols for gates into line with the European symbols. At present a semi-circle is used, but this is to be superseded by a rectangle. Rectangles are used in this book to represent gates. American symbols which vary from manufacturer to manufacturer may be met in industry. Figure 191 shows three inputs to the gate.

The number of inputs used may vary from one gate to another and depends on the purpose for which the gate is required. Similarly the function of the gate depends on the required combination of input signals. The gate is defined according to the combination of input signals required to produce an output signal.

After reading the following material, the reader shall:

20.4 State the logical function of the 'AND' gate.
20.5 Construct a truth table for three input 'AND' gate.
20.6 State the Boolean symbol for 'AND'.

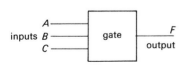

Figure 191 Gating circuit.

In the 'AND' gate the required combination of input signals is as follows:

For the 'AND' gate a logic 1 output is obtained if and only if all the inputs are at logic 1.

This is the logical function provided by the 'AND' gate.

A	B	C	F
0	0	0	0
0	0	1	0
0	1	0	0
0	1	1	0
1	0	0	0
1	0	1	0
1	1	0	0
1	1	1	1

Table 13   Truth table for the 'AND' gate.

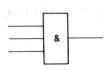

Figure 192   New BS circuit symbol for an 'AND' gate.

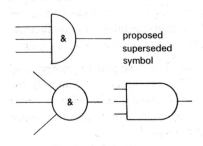

proposed superseded symbol

other symbols in use

Figure 193   Superseded BS circuit symbol for an 'AND' gate.

In Table 13 all of the possible logic combinations are shown. From the logical function it can be seen that the only time a logic output is obtained, is when input A and input B and input C are at logic 1.

A table that shows all possible input combinations together with the respective outputs is called a *truth table*.

Another way of representing the logic AND function is to use the Boolean symbol for 'AND' which is given by '.'. Thus a form of shorthand with a specific meaning is now available for the 'AND' gate.

e.g. $A.B.C = F$

After reading the following material, the reader shall:

20.7   Draw the new BS circuit symbol for an 'AND' gate.
20.8   Recognize the superseded BS circuit symbol for an 'AND' gate.

The new symbol is shown in Figure 192.

The '&' symbol indicates that however many inputs the gate has all of these must be at logic 1 to give a logic 1 output.

The symbol to be superseded is the semi-circular symbol shown in Figure 193. Also shown are other symbols which may be met but do not conform to BS specifications.

Solutions to self-assessment questions (pages 159 and 160)

1   Attenuation.

2   Long.

3   See Figure 188.

4
(a)   Lamp, buzzer.
(b)   Pen recorder, magnetic surface.

5   The lamp goes ON, remains ON, goes OFF, goes ON, goes OFF.

6   10001 positive logic
10001 negative logic
11000 negative logic

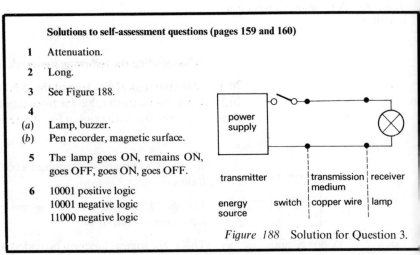

Figure 188   Solution for Question 3.

After reading the following material, the reader shall:

20.9   State the logical function of the 'OR' gate.
20.10   Construct a truth table for a three input 'OR' gate.
20.11   State the Boolean symbol for 'OR'.

A	B	C	F
0	0	0	0
0	0	1	1
0	1	0	1
0	1	1	1
1	0	0	1
1	0	1	1
1	1	0	1
1	1	1	1

Table 14   Truth table for the 'OR' gate.

Another type of gate requires a different combination of input signals to provide an output signal. This is called the 'OR' gate and depends on the following logical function.

For the 'OR' gate a logic 1 output is obtained if and only if any one of the inputs is at logic 1.

Table 14 shows all of the possible logic combinations. From the logical function it can be seen that a logic 1 output is obtained whenever a logic 1 is present on input *A* or input *B* or input *C*.

The representation for this operation that the logic gate performs is called a Boolean symbol. For the 'OR' gate this is written

$$A + B + C = F$$

where the '+' is the Boolean symbol to represent the 'OR' function.

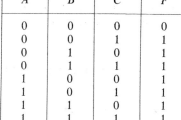

Figure 194   New BS circuit symbol for an 'OR' gate.

After reading the following material, the reader shall:

20.12   Draw the new BS circuit symbol for an 'OR' gate.
20.13   Recognize the superseded BS circuit symbol for the 'OR' gate.

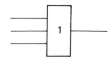

proposed superseded symbol

The new symbol is shown in Figure 194.

The '1' symbol indicates that however many inputs the gate has at least one must be at logic 1 to give a logic 1 output.

The symbol to be superseded is the semi-circular symbol shown in Figure 195. Also shown are other symbols which may be met but do not conform to BS specifications.

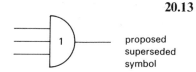

other symbols in use

Figure 195   Superseded BS circuit symbol for an 'OR' gate.

After reading the following material, the reader shall:

20.14   State the logical function of the 'NOT' gate.
20.15   Construct a truth table for a 'NOT' gate.
20.16   State the Boolean symbol for 'NOT'.

This is a gate with one input and one output.

A	F
0	1
1	0

Table 15   Truth table for the 'NOT' gate.

For the 'NOT' gate the output signal is the inverse of the input signal whatever the logic state of the input signal.

Table 15 shows all of the possible logic combinations. Only two possible combinations are shown as only one input exists.

There are a number of words that are used to show this logical function, e.g. inverse, complement and negate.

The representation for this operation that the logical gate performs is called a Boolean symbol. For the 'NOT' gate this is written

$$F = \bar{A}$$

In words, F is the complement of A, or F is the inverse of A, or F is the negation of A.

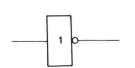

Figure 196   New BS circuit symbol for a 'NOT' gate.

The line over the letter is the Boolean symbol to represent the 'NOT' function.

After reading the following material, the reader shall:

20.17   Draw the new BS circuit symbol for a 'NOT' gate.
20.18   Recognize the superseded BS circuit symbol for a 'NOT' gate.

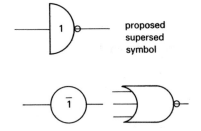

proposed supersed symbol

other symbols in use

Figure 197   Superseded BS circuit symbol for a 'NOT' gate.

The new symbol is shown in Figure 196.

The circle indicates that inversion of the logic signal takes place.

The symbol to be superseded is the semi-circular symbol shown in Figure 197. Also shown are other symbols which may be met but do not conform to BS specifications.

Self-assessment questions

7 Complete the truth tables shown in Table 16 for the one input, the two input and the three input gates.

Table 16 Truth tables for Question 7.

A	B	F
0	0	
0	1	
1	0	
1	1	

'AND' gate

A	B	F
0	0	
0	1	
1	0	
1	1	

'OR' gate

A	F
0	
1	

'NOT' gate

A	B	C	F
0	0	0	
0	0	1	
0	1	0	
0	1	1	
1	0	0	
1	0	1	
1	1	0	
1	1	1	

'AND' gate

A	B	C	F

'OR' gate

8 Using the truth tables completed in Question 7, answer the following questions. Tick the correct alternatives.

(i)   In the 'AND' gate a logic 1 is obtained at the output when:
(a)   ALL inputs are at logic 1
(b)   NO inputs are at logic 1
(c)   ANY one output is at logic 1

(ii)   In the 'OR' gate a logic 1 is obtained at the output when:
(a)   ALL inputs are logic 1
(b)   NO inputs are logic 1
(c)   ANY one input is at logic 1

(iii)   In the 'NOT' gate a logic 1 is obtained at the output when:
(a)   the input is at logic 1
(b)   the input is at logic 0

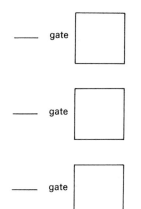

Figure 198 New BS circuit symbols.

9 Draw in the boxes provided in Figure 198 the new BS circuit symbols for the three logic gates, naming each gate in the appropriate space.

10   For each of the three input logic gates of Question 9, write down the respective Boolean representation, if the inputs are *A*, *B* and *C* and the output is *F*, ensuring that each representation is labelled by its respective gate.

11   Write down the Boolean representation for a 'NOT' gate with an input *A* and an output *F*.

Tick the correct alternative.

12   Which switch combination gives the same truth table as the 'AND' gate?   SERIES/PARALLEL

13   Which switch combination gives the same truth table as the 'OR' gate?   SERIES/PARALLEL

14   Mark in the space provided the output signal that would be obtained for each of the pulse train inputs to the gates shown in Figure 199, assuming positive logic is used.

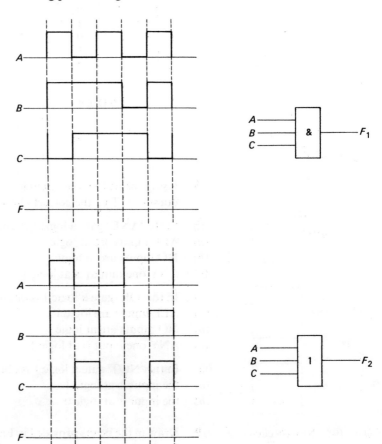

Figure 199   Pulse train inputs and output.

After reading the following material, the reader shall:

21 Understand the action of simple electronic gates.
21.1 Explain the action of a three input diode 'AND' gate.
21.2 Explain the action of a three input diode 'OR' gate.

It has already been explained in topic area Elementary theory of semi-conductors how the semi-conductor diode operates. A very simplified representation is shown in Figure 202.

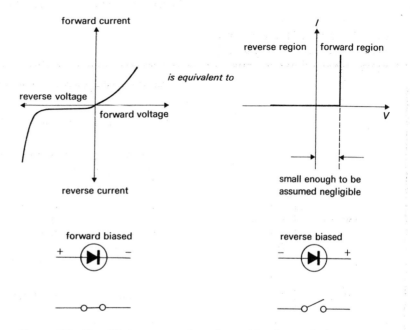

Figure 202 Simplified representation of a semi-conductor diode.

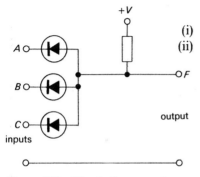

Figure 203 Circuit diagram using three diodes and +V.

This assumes:

(i) The reverse resistance of the diode is infinitely large.
(ii) The forward resistance of the diode is zero and hence the volt drop across a forward biased diode is zero.

Consider the circuit, shown in Figure 203, where three diodes are connected via a resistor to a positive potential V volts. The output is taken from the junction of the diodes and the resistor. If the inputs (using positive logic) have a low positive potential for '0' state and a higher positive potential for the '1' state, then when the three inputs are all at the '0' state all of the diodes are *forward biased and hence switched on*. This assumes that the potential level for the '0' state is less than V volts.

inputs			output
A	*B*	*C*	*F*
0	0	0	0
0	0	1	0
0	1	0	0
0	1	1	0
1	0	0	0
1	0	1	0
1	1	0	0
1	1	1	1

Table 18   Truth table using positive logic for circuit shown in figure 203.

The output is now at the '0' state, because the diodes are on and the volt drop across them is therefore zero.

If one input is set to the '1' state (potential level just above *V* volts) the output state is not affected because only one diode is switched off.

If all the inputs are set to the '1' state all the diodes are *reversed biased and hence switched off.* Thus no current can flow through the resistor (assuming the output feeds into an infinitely high resistance) and the output has a potential *V* volts and is in the '1' state.

The truth table for this is shown in Table 18.

7

Table 17   Solution for Question 7.

input		'AND' output	'OR' output
A	*B*	*F*	*F*
0	0	0	0
0	1	0	1
1	0	0	1
1	1	1	1

A	*F*
0	1
1	0

inputs			'AND' output	'OR' output
A	*B*	*C*	*F*	*F*
0	0	0	0	0
0	0	1	0	1
0	1	0	0	1
0	1	1	0	1
1	0	0	0	1
1	0	1	0	1
1	1	0	0	1
1	1	1	1	1

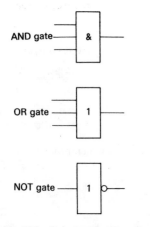

AND gate

OR gate

NOT gate

Figure 200   Solution for Question 9.

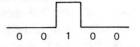

0   0   1   0   0

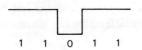

1   1   0   1   1

Figure 201   Solution for Question 14.

8   (i) *a*   (ii) *a* and *c*   (iii) *b*

9   See Figure 200.

10   AND $A.B.C. = F$
OR $A + B + C = F$

11   NOT $F = \bar{A}$

12   SERIES.

13   PARALLEL.

14   See Figure 201.

inputs			output
A	*B*	*C*	*F*
1	1	1	0
1	1	0	1
1	0	1	1
1	0	0	1
0	1	1	1
0	1	0	1
0	0	1	1
0	0	0	1

Table 19   Truth table using negative logic for circuit shown in figure 203.

This is the truth table for the 'AND' gate when positive logic is used.

If negative logic is used the more positive potential level is called the '0' state and the less positive potential is called the '1' state. Thus if the same potential levels are used with negative logic, then the truth table shown in Table 19 is obtained.

This is the truth table for the 'OR' gate when negative logic is used.

Consider the circuit diagram, shown in Figure 204, where the diodes are connected the other way round, and the potential is changed to a small negative potential.

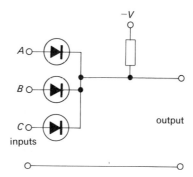

Figure 204   Circuit diagram using three diodes and − V.

inputs			output
A	*B*	*C*	*F*
0	0	0	0
0	0	1	1
0	1	0	1
0	1	1	1
1	0	0	1
1	0	1	1
1	1	0	1
1	1	1	1

Table 20   Truth table using positive logic for circuit shown in figure 204.

Using the positive logic system previously described consider what happens when all the inputs are set to the '0' state.

All the diodes are forward biased and conduct (and no voltage is dropped across the diodes) causing the potential at the output to be a small negative potential. This is taken to be the '0' state. If an increase of potential at any one input occurs the diode is still forward biased and the output follows this increase of potential. Hence any input in the '1' state produces an output in the '1' state. The reason for using diodes is to prevent any feedback of an input in the '1' state to the input that is in the '0' state. The truth table for this circuit is shown in Table 20.

This is the truth table for the 'OR' gate when positive logic is used.

When negative logic is used for the same potential levels described the truth table shown in Table 21 is obtained.

inputs			output
A	*B*	*C*	*F*
1	1	1	1
1	1	0	0
1	0	1	0
1	0	0	0
0	1	1	0
0	1	0	0
0	0	1	0
0	0	0	0

Table 21   Truth table using negative logic for circuit shown in figure 204.

Thus when the circuit is used with negative logic the circuit is called the 'AND' gate.

After reading the following material, the reader shall:

21.3 Explain the action of a transistor when used as a switch.

21.4 Explain the action of a transistor 'NOT' gate.

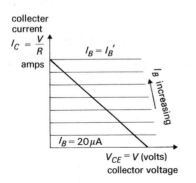

Figure 205 Simplified transistor characteristics.

As was explained in topic area Elementary theory of semiconductors, the magnitude of current flowing in the base of a common emitter transistor circuit controls the current flowing from the collector to the emitter. A very simplified form of transistor output characteristics is shown in Figure 205.

The load line is drawn assuming a fixed value of collector resistance R and a supply voltage V.

For $I_B = 0$ the collector voltage $= V$

$$\text{collector current} = 0$$

For $I_B = I'_B$ then collector voltage $= 0$

$$\text{collector current} = \frac{V}{R} \text{ amps}$$

(In practice a small leakage current flows if the base potential is made slightly negative, but this has been ignored.)

This means the transistor can be assumed to be equivalent to the switch shown in Figure 206.

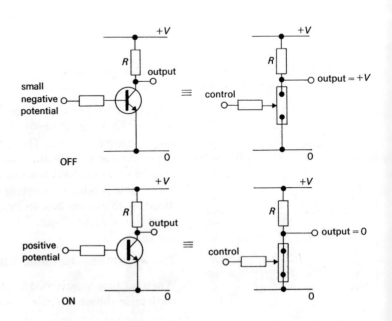

Figure 206 Transistor as a switch.

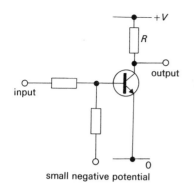

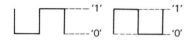

Figure 207 The transistor used as a 'NOT' gate.

input	output
A	*F*
0	1
1	0

Table 22 'NOT' gate truth table using positive logic.

input	output
A	*F*
1	0
0	1

Table 23 'NOT' gate truth table using negative logic.

For a small negative applied potential the transistor is OFF.

For a large enough positive applied potential the transistor is ON.

This positive potential must be large enough to ensure enough base current flows, hence causing the transistor to *saturate* and be switched ON.

When the transistor is OFF the output voltage is *V* volts. When the transistor is ON the output voltage is zero.

A similar explanation can be used to describe the operation of the p–n–p transistor by applying the correct potentials to the circuit.

In the circuit shown in Figure 207, the transistor is held off by making the supply to the base a small negative potential. If positive logic is used what happens when the train of pulses shown in Figure 207 is applied to the input? Zero potential represents the '0' state and a positive potential represents the '1' state. With no pulses applied the transistor is switched OFF. Applying the pulse at the '0' state does not affect the transistor which remains OFF.

No current flows through the resistor *R* and hence there is no potential drop across it.

Therefore the output is the supply potential *V* volts and at logic state '1'.

Applying the pulse at the '1' state switches the transistor ON, bringing the output potential to zero volts. Therefore the output is at logic state '0'.

The truth table for this, shown in Table 22, is the one for the 'NOT' gate. If the logic is changed to negative logic the truth table is the one for the 'NOT' gate, shown in Table 23. This again assumes voltage levels are not changed.

At present one of the most popular types of logic is known as transistor transistor logic (TTL) which may be connected to use positive logic as follows:

'1' state corresponds to a voltage level between 2·4 and 5·25 V.
'0' state corresponds to a voltage level between 0 and 0·4 V.

Self-assessment questions

15 In the following sentences delete the incorrect word.

When a diode is forward biased it can be considered similar to a switch that is ON/OFF.
This assumes that the forward resistance of the diode is LOW/HIGH.

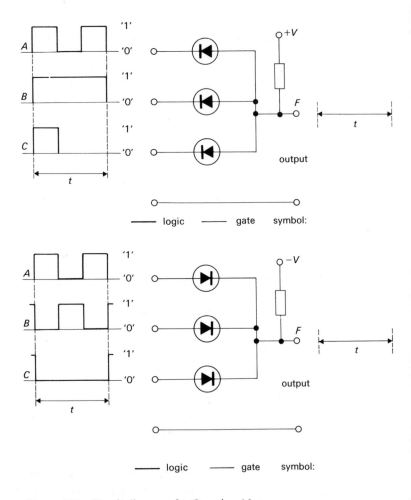

Figure 208   Circuit diagrams for Question 16.

16   For the circuit diagrams shown in Figure 208 complete the output waveform.
Also write in the space provided the logic used, the name of the gate, and the new BS circuit symbol for this gate.

17   Complete the following sentences by deleting the incorrent word.

An n–p–n transistor is connected in the common emitter mode and used as a high speed switch. When the base current flowing is HIGH/ZERO there is no collector current flowing and the transistor is said to be ON/OFF.
When the base current flowing is large enough to cause the transistor to CUT OFF/SATURATE the transistor is said to be ON/OFF.

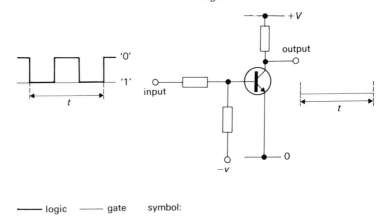

Figure 209 shows the circuit diagram with waveform labeled '0' and '1' over time *t*, an input, and a transistor circuit with +V, output, 0, and −V, and the label "logic —— gate symbol:"

Figure 209 Circuit diagram for Question 18.

18 For the circuit diagram shown in Figure 209 complete the output waveform. Also write in the space provided the logic used, the name of the gate and the new BS circuit symbol for this gate.

Post test – Logic elements and circuits

For the following questions tick the correct alternative. There may be more than one correct alternative in each question.

19 In the 'AND' gate a logic 0 output is obtained when:

(*a*) all inputs are at logic 0
(*b*) no inputs are at logic 0
(*c*) any one input is at logic 0

20 In the 'AND' gate a logic 1 output is obtained when:

(*a*) all inputs are at logic 1
(*b*) no inputs are at logic 1
(*c*) any one input is at logic 1

21 In the 'OR' gate a logic 0 output is obtained when:

(*a*) all inputs are at logic 0
(*b*) no inputs are at logic 0
(*c*) any one input is at logic 0

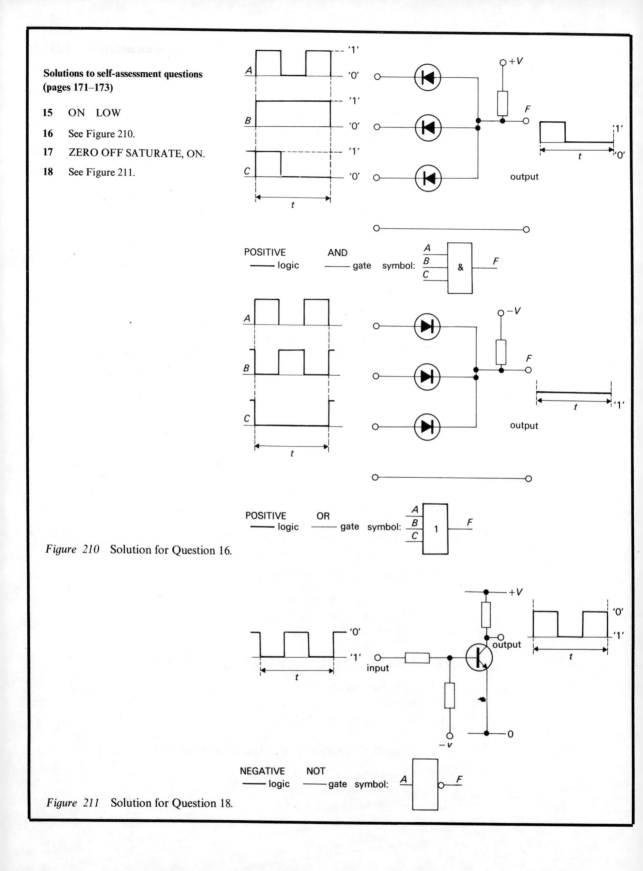

Solutions to self-assessment questions (pages 171–173)

15 ON   LOW

16 See Figure 210.

17 ZERO OFF SATURATE, ON.

18 See Figure 211.

POSITIVE — logic   AND — gate   symbol:

Figure 210 Solution for Question 16.

Figure 211 Solution for Question 18.

22 In the 'OR' gate a logic 1 output is obtained when:

(*a*) all inputs are at logic 1
(*b*) no inputs are at logic 1
(*c*) any one input is at logic 1

23 Complete the following sentences by deleting the incorrect word.

(*a*) A logic 1 is obtained from an 'OR' gate if and only if ANY ONE/ALL of the inputs IS/ARE at logic 1.
(*b*) A logic 1 is obtained from an 'AND' gate if and only if ANY ONE/ALL of the inputs IS/ARE at logic 1.
(*c*) The output signal of the 'NOT' gate is the REPLICA/INVERSE of the input signal whatever the logic state of the input signal.

24 Complete the columns in Table 24 for a three input 'OR' and 'AND' gate, and a 'NOT' gate.

Gate	New B.S. Symbol	Boolean representation
'OR'		
'AND'		
'NOT'		

Table 24 Table for Question 24.

25 The truth table shown in Figure 212 represents the possible input conditions for the logic diagram shown in the same figure, for a particular set of conditions.
Which set of input conditions gives a logic 0 output? Tick the correct row.

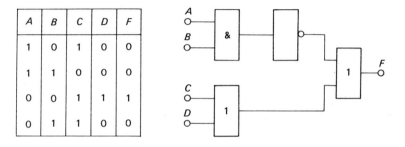

A	B	C	D	F
1	0	1	0	0
1	1	0	0	0
0	0	1	1	1
0	1	1	0	0

Figure 212 Truth table and diagram for Question 25.

26   In the following sentences delete the incorrect words. Assume positive logic is used throughout.

(*a*)   A three input diode 'OR' gate has all inputs at logic 0. All the diodes are FORWARD/REVERSE biased and ON/OFF and hence the output is at LOGIC 0/LOGIC 1. If only one of the inputs is at logic 1, the output is at LOGIC 0/LOGIC 1. The purpose of the diodes is to prevent feedback to the inputs at LOGIC 0/LOGIC 1.

(*b*)   A three input diode 'AND' gate has all inputs at logic 0. All the diodes are FORWARD/REVERSE biased and ON/OFF and hence the output is at LOGIC 0/LOGIC 1. If only one of the inputs is at logic 1, the output is at LOGIC 1/LOGIC 0. All of the inputs must be at LOGIC 0/LOGIC 1 before the output changes state.

(*c*)   In the transistor 'NOT' gate if the input is at logic 0 the transistor is held ON/OFF, and the output is at LOGIC 0/LOGIC 1.
If the input is at logic 1, the transistor is held ON/OFF and the output is at LOGIC 0/LOGIC 1.

The circuit diagrams are given in Figure 213.

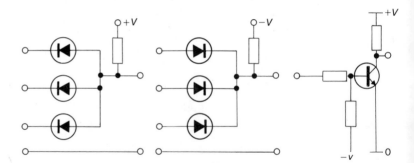

Figure 213   Circuit diagrams for Question 26.

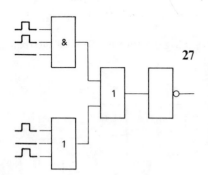

Figure 214   Logic diagram for Question 27.

27   Name the logic gates used in the logic diagram shown in Figure 214. Write the answers next to the appropriate gate. At the output of each gate mark on the correct pulse, and hence write the correct logic state at the output.

Gate	B.S. 3939 proposed symbol	Boolean representation
'OR'	⊐1⊢	+
'AND'	⊐&⊢	•
'NOT'	⊐1○⊢	−

Table 25   Solution for Question 24.

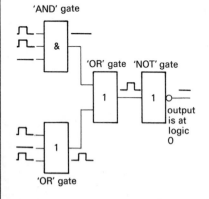

Figure 215   Solution for Question 27.

Solutions to post test (pages 173 and 175–6)

19   *a, c*

20   *a*

21   *a*

22   *a, c*

23
(*a*)   anyone   is
(*b*)   all   are
(*c*)   inverse

24   See Table 25.

25   *A   B   C   D   F*
1   1   0   0   0   correct row

26   forward,   on,   logic 0,   logic 1,   logic 0
(*b*)   forward,   on,   logic 0,   logic 0,   logic 1
(*c*)   OFF,   logic 1,   ON,   logic 0

27   See Figure 215.

Appendix A Units, symbols and decimal prefixes

Units and symbols

quantity	unit name	unit symbol
resistance	ohm	Ω
capacitance	farad	F
inductance	henry	H
voltage	volt	V
current	ampere	A
power	watt	W
energy	joule	J
charge	coulomb	C

Decimal prefixes

prefix	name	quantity	standard form
M	mega	1 000 000	10^6
k	kilo	1 000	10^3
m	milli	1/1 000	10^{-3}
μ	micro	1/1 000 000	10^{-6}
n	nano	1/1 000 000 000	10^{-9}
p	pico	1/1 000 000 000 000	10^{-12}

Appendix B Circuit symbols common in electronic circuits

———	single conductor
earth symbol	earth connection
battery symbol	battery consisting of 2 single cells (positive terminal is the long thin line)
switch symbol	single pole, single throw switch
switch symbol	single pole, double throw switch
push button symbol	push button switch (push to make)
push button symbol	push button switch (push to break)
microphone symbol	microphone
earphone symbol	earphone
loudspeaker symbol	loudspeaker
resistor symbol	fixed value resistor
variable resistor symbol	variable resistor with control knob
variable resistor symbol	variable resistor with preset adjustment
capacitor symbol	fixed value capacitor
electrolytic capacitor symbol	electrolytic capacitor (polarized)
variable capacitor symbol	variable capacitor with control knob
variable capacitor symbol	variable capacitor with preset adjustment
inductor symbol	winding of an inductor, coil or transformer
meter symbol (A)	meter (various types are designated by letter or sign inside the circle e.g. A-Ammeter)

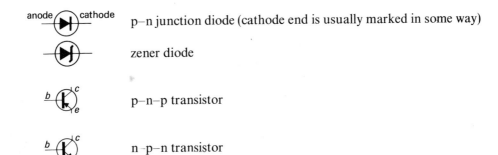

p–n junction diode (cathode end is usually marked in some way)

zener diode

p–n–p transistor

n–p–n transistor

Appendix C Resistor colour code

Most carbon composition resistors are marked with coloured bands which indicate the ohmic value and tolerance of the resistor. The data below explains how to interpret the ohmic value and tolerance from the coloured bands.

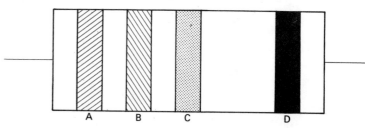

Band A—digit
Band B—digit
Band C—multiplier (the number of zeros in the ohmic value)
Band D—tolerance

Digit and multiplier band

colour	numerical value
black	0
brown	1
red	2
orange	3
yellow	4
green	5
blue	6
violet	7
grey	8
white	9

Tolerance band

no band	$\mp 20\%$
silver	$\mp 10\%$
gold	$\mp 5\%$
red	$\mp 2\%$
brown	$\mp 1\%$

Gold and silver are also used as multipliers. They represent 0·1 and 0·01 times the value respectively. A fifth band is sometimes used, coloured salmon pink to indicate a high stability resistor.

Examples of resistor colour code

band A	band B	band C	band D	ohmic value
red	red	brown	gold	$220\,\Omega \mp 5\%$
yellow	violet	orange	silver	$47\,000\,\Omega \mp 10\%$
				i.e. $47\,k\Omega \mp 10\%$
brown	black	green	no band	$1\,000\,000\,\Omega \mp 20\%$
				i.e. $1\,M\Omega \mp 20\%$
orange	white	gold	gold	$39 \times 0{\cdot}1\,\Omega \mp 5\%$
				i.e. $3{\cdot}9\,\Omega \mp 5\%$

Appendix D Capacitor colour code

Two capacitor colour codes exist, one for ceramic and polyester capacitors and one for tantalum capacitors. Both use the same colour code as resistors but have different multiplier codes.

Ceramic and polyester types

colour	numerical value (tens and units)	multiplier	tolerance
black	0	$\times 1\,pF$	$\mp 20\%$
brown	1	$\times 10\,pF$	$\mp 1\%$
red	2	$\times 100\,pF$	$\mp 2\%$
orange	3	$\times 1\,000\,pF$	$\mp 2.5\%$
yellow	4	$\times 10\,000\,pF$	—
green	5	$\times 1\,000\,000\,pF$	$\mp 5\%$
blue	6		—
violet	7		—
grey	8		—
white	9		$\mp 10\%$

ceramic types (4 dot colour code)

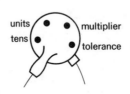
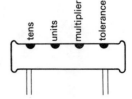

polyester type

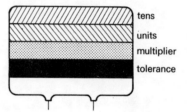

Tantalum types

colour	numerical value (tens and units)	multiplier	voltage
black	0	× 1 F	10V
brown	1	× 10 F	
red	2	× 100 F	
orange	3		
yellow	4		6·3 V
green	5		16 V
blue	6		20 V
violet	7		
grey	8	× 0·01 F	25 v
white	9	× 0·1 F	3 v
pink			35 v

Tantalum types

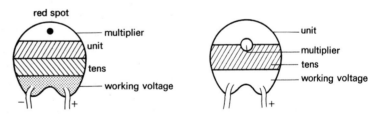

When viewed with spot showing as above the positive lead is as marked.

Appendix E Testing transistors

Many commercial instruments are available to test transistors for input and output resistance and current gain. In most cases however, it is usually only necessary to detect if a transistor is suspect or unserviceable. An AVO meter (set on the ohms range) is usually quite adequate to test this. Three tests are required:

1 The resistance between base and emitter should be *low* ($\simeq 1 \, k\Omega$) with the AVO test leads one way round but *high* the other way round.

2 The resistance between base and collector should be *low* with the AVO test leads one way round but *high* the other way round.

3 The resistance between emitter and collector should be *high* with the AVO test leads either way round.

The above three tests are true for both p–n–p and n–p–n transistors.

Note: It may be unwise to use the high ohms range on the AVO because the battery voltage on this range could damage the transistor.

Unknown transistors may also be sorted into the two types, p–n–p and n–p–n, as follows:
With the *red* lead of the AVO to the *base* and the *black* lead to the *emitter* a LOW reading is obtained for a p–n–p transistor and a HIGH reading for a n–p–n transistor. This appears to contradict the p–n junction theory but it must be remembered that an AVO delivers a *positive* potential from its *negative* terminal, and a *negative* potential from its *positive* terminal when set on the ohms range.